Simulation Conceptual Modeling

by

Jeffrey S. Strickland

Simulation Conceptual Modeling

Copyright 2011 by Jeffrey S. Strickland. All rights Reserved

ISBN 978-1-105-18162-7

 www.simulation-educators.com

"Global Leaders in Training"

Published by Lulu, Inc.

Wikipedia® is a registered trademark of the Wikimedia Foundation, Inc., a non-profit organization.

All pictures, unless otherwise cited, are taken from the Wikimedia Commons of the Wikimedia Foundation, and are either public domain or used under the terms of the GNU Free Documentation License or the Creative Commons Attribution/Share-Alike License. Public domain pictures have been place in public domain by the authors or their copyrights have expired.

Acknowledgement

This book is dedicated to Mariah Strickland and Evie Strickland for all their loving support.

A special thanks is extended to Laurie Parham Strickland—loving wife, nurturing mother, and loyal friend.

Foreword

Simulation Conceptual Modeling is the result of an apparent hole in the body of simulation knowledge. A colleague came to me a few months ago and asked for help with conceptual modeling for a simulation framework. As I began to recommend sources, I began to realize that there no definitive work on the subject.

I would like to extend a special thanks to Adam Sand and Chuck Piersall of the Missile Defense Agency Modeling and Simulation Directorate for inspiring this book.

Table of Contents

ACKNOWLEDGEMENT I
FOREWORD II
TABLE OF CONTENTS III
PREFACE XI

CHAPTER 1. MODELING AND SIMULATION: CONCEPTUAL MODELING OVERVIEW 1

    INTRODUCTION 1
    SIMULATION CONCEPTUAL MODEL COMPONENTS 4
    WHAT INFORMATION SHOULD A SIMULATION CONCEPTUAL MODEL INCLUDE? 8
    NOTES 9

CHAPTER 2. SIMULATION CONCEPTUAL MODEL DEVELOPMENT 11

    DEVELOPMENT PROCESS 11
        1. Collect Authoritative Information. 11
        2. Decompose the Mission Space 14
        4. Identify Relationships 20
        5. Assess and Record 20
    DEVELOPMENT CONSIDERATIONS 21
        Reality Abstraction 21
        Problem Identification 22
    NOTES 22

CHAPTER 3. CONCEPTUAL MODELING TECHNIQUES OVERVIEW 23

    Structured System Analysis and Design Method 23
    Soft System Method 24
    Data Flow Modeling 24
    Entity Relationship Modeling 24
    Event-Driven Process Chain 25
    Joint Application Development 25
    Place/ Transition Net 26
    State Transition Modeling 26
    Object Role Modeling 26
    Unified Modeling Language 27
    NOTES 27

CHAPTER 4. STRUCTURED SYSTEMS ANALYSIS AND DESIGN METHOD 29

OVERVIEW 29
HISTORY 30
SSADM TECHNIQUES 31
STAGES 31
Stage 0 - Feasibility study 32
Stage 1 - Investigation of the current environment 32
Stage 2 - Business system options 34
Stage 3 - Requirements specification 35
Stage 4 - Technical system options 36
Stage 5 - Logical design 36
Stage 6 - Physical design 37
Advantages and disadvantages 38
NOTES 38

CHAPTER 5. SOFT SYSTEMS METHODOLOGY 41

OVERVIEW 41
HISTORY 42
THE 7-STAGE APPROACH OF SSM 44
CATWOE 44
Conceptual Models of Human Activity Systems 45
Outcomes and applications of SSM 47
SSM for Information Systems Analysis and Design 48
NOTES 50

CHAPTER 6. SOFT MODEL ANALYSIS FOR DISCRETE EVENT SIMULATION (DES) 53

OVERVIEW 53
ANALYSIS PREPARATION 54
(Step 1) Determine Prerequisites 54
(Step 2) Select Analyst Subject Matter Experts (SMEs) .. 55
(Step 3) Assemble Conceptual Model Documentation ... 56
(Step 4) Select Input and Output Variables 57
(Step 5) Develop a Soft Model 57
(Step 6) Build the Discrete Event Model 58
(Step 7) Build the DOE and Make Simulation Runs 58
(Step 8) Apply the Validation Technique 59
(STEP 9) PREPARE THE REPORT 63

NOTES 65

CHAPTER 7. STRUCTURED ANALYSIS 69

OBJECTIVES OF STRUCTURED ANALYSIS 69
HISTORY 70
STRUCTURED ANALYSIS TOPICS 72
    Single abstraction mechanism 72
    Approach 73
    Context diagram 74
    Data dictionary 75
    Data Flow Diagrams 77
    Structure Chart 78
    Structured Design 79
    Structured query language 80
CRITICISMS 80
NOTES 81

CHAPTER 8. DATA FLOW DIAGRAM 85

OVERVIEW 85
ADVANTAGES OF DFD 87
NOTATIONS OF DFDS 88
    Data Flow 88
    Processes 88
    Data Stores 88
    External Entities 89
    Resource store 89
STEPS 89
    Types of DFD 89
    Rules of DFD 90
CONTEXT DIAGRAMS 91
LEVEL 1 DIAGRAMS 91
RESOURCE FLOW ANALYSIS 93
DOCUMENT FLOW ANALYSIS 93
ORGANIZATIONAL STRUCTURE ANALYSIS 93
NOTES 94

CHAPTER 9. JACKSON STRUCTURED PROGRAMMING 95

INTRODUCTION 95
STRUCTURAL EQUIVALENT 96
THE METHOD 98

A WORKED EXAMPLE 100
NOTES 105

CHAPTER 10. ENTITY RELATIONSHIP MODELING 109

ENTITY-RELATIONSHIP MODEL 110
OVERVIEW 111
RELATIONSHIPS, ROLES AND CARDINALITIES 115
   Relationship Names *115*
   Role naming *116*
   Cardinalities *116*
SEMANTIC MODELING 117
DIAGRAMMING CONVENTIONS 118
   Crow's Foot Notation *121*
ER DIAGRAMMING TOOLS 122
NOTES 123

CHAPTER 11. EVENT-DRIVEN PROCESS CHAIN 127

EVENT-DRIVEN PROCESS CHAIN 127
OVERVIEW 129
ELEMENTS OF AN EVENT-DRIVEN PROCESS CHAIN 130
EXAMPLE 134
META-MODEL OF EPC 134
NOTES 134

CHAPTER 12. JOINT APPLICATION DEVELOPMENT 137

JOINT APPLICATION DESIGN 137
   Origin *138*
   Key participants *138*
   9 Key Steps *139*
   Advantages *142*
   Challenges *143*
NOTES 144

CHAPTER 13. DYNAMIC SYSTEMS DEVELOPMENT METHOD 145

DSDM AND THE DSDM CONSORTIUM: ORIGINS 146
DSDM ATERN 147
OVERVIEW OF DSDM ATERN 148
THE DSDM ATERN APPROACH 148
   Principles *148*
   Prerequisites for using DSDM *151*

OVERVIEW OF DSDM VERSION 4.2 151
DSDM V4.2 PROJECT LIFE-CYCLE 152
   Overview: three phases of DSDM V4.2 .... *152*
   Four stages of the DSDM V4.2 Project life-cycle *155*
DSDM V4.2 FUNCTIONAL MODEL ITERATION 161
   Meta-data model *161*
   Process-data model *165*
   Notes *168*

CHAPTER 14. PLACE/ TRANSITION NET 171

PETRI NET BASICS 172
FORMAL DEFINITION AND BASIC TERMINOLOGY 173
   Syntax *173*
   Execution semantics *174*
VARIATIONS ON THE DEFINITION 175
FORMULATION IN TERMS OF VECTORS AND MATRICES 175
MATHEMATICAL PROPERTIES OF PETRI NETS 177
   Reachability *177*
   Liveness *178*
   Boundedness *179*
DISCRETE, CONTINUOUS, AND HYBRID PETRI NETS 181
EXTENSIONS 181
RESTRICTIONS 184
OTHER MODELS OF CONCURRENCY 186
APPLICATION AREAS 186
NOTES 186

CHAPTER 15. STATE TRANSITION MODELING 191

STATE DIAGRAM 191
OVERVIEW 192
DIRECTED GRAPH 192
   Example: DFA, NFA, GNFA, or Moore machine *194*
   Example: Mealy machine *194*
HAREL STATECHART 195
ALTERNATIVE SEMANTICS 196
STATE DIAGRAMS VERSUS FLOWCHARTS 196
OTHER EXTENSIONS 197
NOTES 198

CHAPTER 16. OBJECT-ROLE MODELING 201

Overview 201
History 202
    ORM2 .... *203*
Object role modeling topics 205
    Graphic notation *205*
    The conceptual schema design procedure *206*
Tools 207
    DogmaModeler *207*
    VisioModeler *208*
    Visio for Enterprise Architects (VEA) *209*
    CaseTalk *210*
    Infagon *211*
    Other tools *212*
    NORMA *212*
    NORMA Project status *212*
Notes 213

CHAPTER 17. UNIFIED MODELING LANGUAGE 215

Overview 215
History 216
    Before UML 1.x *216*
    UML 1.x *218*
    UML 2.x *219*
Topics 219
    Software development methods *219*
    Modeling *220*
    Diagrams overview *220*
    Meta modeling *230*
Criticisms 231
UML modeling tools 234
Notes 235

CHAPTER 18. DODAF 237

DoDAF overview 237
Representation 237

CHAPTER 19. TECHNIQUE EVALUATION AND SELECTION 247

Considering Affecting Factors *247*
Considering Affected Variables *248*
Evaluation Criteria for a Simulation Conceptual Model 249

A MULTI-PERSPECTIVE FRAMEWORK FOR EVALUATION 250
   The Economic Perspective..*251*
   The Engineering Perspective..*252*
   The Deployment Perspective..*253*
   The Epistemological Perspective...*254*
NOTES 257

Preface

If someone were to ask me about verification and validation (V&V) of simulations—people have accused me of being somewhat of an expert in V&V—I would imagine the following conversation.

ME: "Did you get the requirements right?"

DEVELOPER: "Yes."

ME: "Did you get the code right?"

DEVELOPER: "Yes."

ME: "Let me see the conceptual model."

DEVELOPER: "We don't have one."

ME: "Then, I do not believe you."

I have done more than imagine this conversation; I have had it on numerous occasions. Yet, we do very little to help the developer build a conceptual model. There is lots of help for requirements, prototyping, designing, coding, testing, etc., but little in conceptual modeling.

Many would ask, "Is conceptual modeling really that important?' The answer depends partly on the "kind" of simulation you are building. Simulations range from representations for engineering a single part for a system or component, to a federation of simulations representing a complex business process or weapons system. A simulation conceptual model is frequently described as the bridge between the Developer and the User. In the former example, the developer and user may speak the same language. In the latter example, this is unlikely, so the simulation conceptual modeling may be essential.

This question might be answered better if we look at the purposes of simulation conceptual models.

- as a basis for assessment of simulation appropriateness for a particular application
- as a context for results validation
- as a foundation for design of software and other components for new and modified simulations
- as a basis for effective and efficient communication about the simulation and its capabilities among Users, Developers, those involved in simulation-related assessments, and others
- as a tool for enhancing understanding of modeling and simulation (M&S) requirements and their implications for simulation capabilities and costs
- as an important aspect of simulation design/implementation verification
- to facilitate reuse of simulation components in simulation development and evolution

The simulation conceptual model provides a rational basis for judgment about the appropriateness of a simulation for use in situations that are not explicitly tested. It provides a context for results validation so that one has a basis for judgment about acceptability of interpolation or extrapolation of simulation results relative to validation referent data. During simulation development or modification, it is a means by which M&S requirements can be transformed into simulation specifications that then drive simulation design.

The effort required to develop a quality simulation conceptual model is justified by the importance of correctly using a simulation appropriate for its intended uses. When a simulation is involved in critical decisions, whether in support of planning, analysis, design, operation, or training, a quality simulation conceptual model increases the likelihood that the simulation will be used correctly and that appropriate use is made of simulation results.

This book explores several system analysis methods and conceptual modeling techniques. We also discuss appropriate tool that may be used to assist with conceptual modeling. Moreover, we also discuss how to evaluate the quality of a conceptual model.

Other Books by the Author

Weird Scientists – the Creators of Quantum Physics. Copyright 2011 by Jeffrey S. Strickland. Lulu.com. ISBN 978-1-257-97624-9

Men of Manhattan: Creators of the Nuclear Age. Copyright © by Jeffrey S. Strickland. Lulu.com. ISBN 978-1-257-76188-3

Albert Einstein: "Nobody expected me to lay golden eggs". Copyright © by Jeffrey S. Strickland. Lulu.com. ISBN 978-1-257-86014-2

Quantum Phaith. Copyright © 2011 by Jeffrey S. Strickland. Lulu.com. ISBN 978-1-257-64561-9

Using Math to Defeat the Enemy: Combat Modeling for Simulation. Copyright © by Jeffrey S. Strickland. Lulu.com. 978-1-257-83225-5

Mathematical Modeling of Warfare and Combat Phenomenon. Copyright © 2011 by Jeffrey S. Strickland. Lulu.com. ISBN 978-1-4583-9255-8

Missile Flight Simulation – Surface-to-Air Missiles. Copyright © 2010 by Jeffrey S. Strickland. Lulu.com. ISBN 978-0-557-88553-4

Discrete Event Simulation using ExtendSim 8. Copyright © 2010 by Jeffrey S. Strickland. Lulu.com. ISBN 978-0-557-72821-3

Fundamentals of Combat Modeling. Copyright © 2010 by Jeffrey S. Strickland. Lulu.com. ISBN 978-1-257-00583-3

Systems Engineering Processes and Practice. Copyright © 2010 by Jeffrey S. Strickland. Lulu.com. ISBN 978-1-257-09273-4

Chapter 1. Modeling and Simulation: Conceptual Modeling Overview

Introduction

"The conceptual model for a simulation addresses the simulation's context, how it will satisfy its requirements, and how its entities and processes will be represented. The conceptual model is key to (1) assessing a simulation's validity for any situation not explicitly tested and (2) determining the appropriateness of a simulation (or its parts) for reuse or use with other simulations in a distributed simulation. There are no widely accepted approaches for decomposing the representation of the simulation subject into the entities and processes of a simulation's conceptual model, for abstracting such representation from available information about the subject, or for describing and documenting the simulation's conceptual model." [1]

- Dale K. Pace

A **simulation conceptual model** is the simulation developer's way of translating modeling requirements (i.e., what is to be represented by the simulation) into a detailed design framework (i.e., how it is to be done), from which the software, hardware, networks (in the case of distributed simulation), and systems/equipment that will make up the simulation can be built. Since simulations vary from stand-alone to federated digital and hard-ware-in-the-loop (HWIL), the complexity of conceptual models—and the tools used to build them—vary as well.

The simulation conceptual model is the generic idea for the simulation to support its full spectrum of applications. For data-driven simulations, it is possible either to construct the simulation with data embedded or to construct the simulation so that it draws data from the inputs for a particular application of the simulation.

Different conceptual models would be used for each approach, as illustrated in the following example.

> Example: If one wants to explore potential consequences of changing the power of a radar represented in the simulation, that parameter could be set by an input value and a single simulation conceptual model would suffice for multiple applications, each of which varied the radar's power. The radar's power could also be made an embedded parameter of the simulation element describing the radar, in which case a separate conceptual model would be required for each version of the radar considered (i.e., those with greater or lesser power).

A **primary function** of the simulation conceptual model is to serve as the mechanism by which simulation requirements are transformed into detailed simulation specifications (and associated simulation design) which fully satisfy the requirements. This transformation is easiest and most reliable if both the requirements and the specifications can be expressed in the same descriptive formalism, because every translation from one descriptive formalism to another introduces an additional source of potential error, even for mundane transformations. Errors may result from something as simple as failure to translate units, which caused failure of NASA's Mars Climate Orbiter in September 1999 [2].

A conceptual model's **primary objective** is to convey the fundamental principles and basic functionality of the system in which it represents. In addition, a conceptual model must be developed in such a way as to provide an easily understood system interpretation for the models users. A conceptual model, when implemented properly, should satisfy five fundamental objectives [3].

- Enhance an individual's understanding of the representative system
- Facilitate efficient conveyance of system details between stakeholders[1]

- Provide a point of reference for system designers to extract system specifications
- Document the system for future reference and provide a means for collaboration
- Provide a rational and factual basis for assessment of simulation application appropriateness

The conceptual model plays an important role in the overall system development life cycle[2]. Figure 1-1 [4] below, depicts the role of the conceptual model in a typical system development scheme.

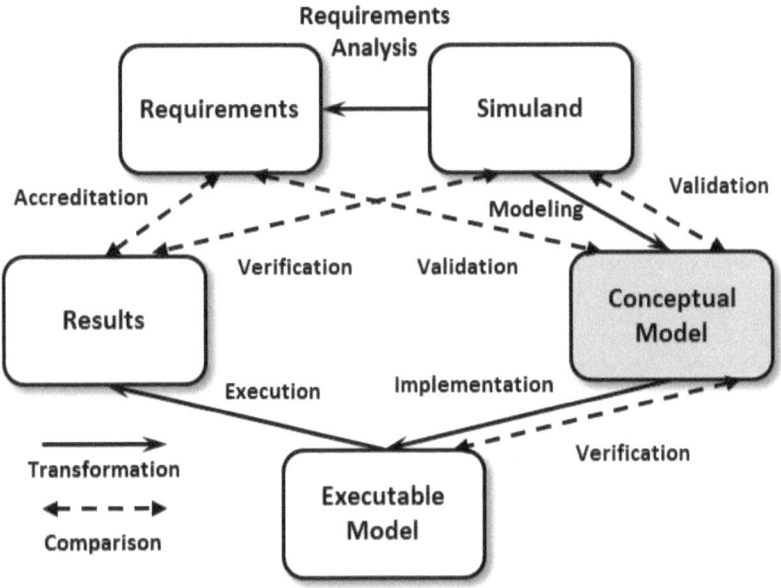

Modeling and Simulation Fundamentals: Theoretical Underpinnings and Practical Domains
By John A. Sokolowski, Catherine M. Banks

Figure 1-1. Conceptual Modeling in the system development life cycle

It is clear that if the conceptual model is not fully developed, the execution of fundamental system properties may not be implemented properly, giving way to future problems or system shortfalls. These failures do occur in the industry and have been linked to; lack of user input, incomplete or unclear requirements, and changing requirements. Those weak links in the system design and development process can be traced to improper execution of the fundamental objectives of conceptual modeling. The importance of conceptual modeling is evident when such systemic failures are mitigated by thorough system development and adherence to proven development objectives/techniques.

Simulation Conceptual Model Components

A simulation's conceptual model consists of the simulation context, the simulation concept (with its mission space and simulation space aspects), and simulation elements. We briefly discuss each of these below. The relationships among these are illustrated by Figure 1-2.

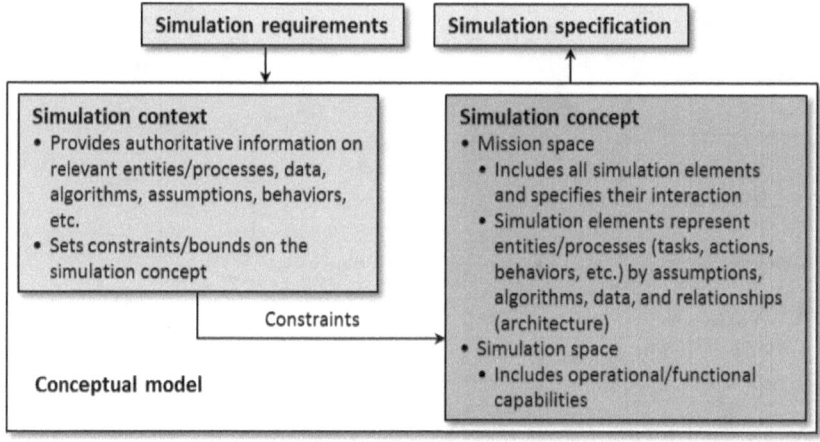

Figure 1-2. Conceptual Model Components

1. The **simulation context** provides "authoritative" information about the domain that the simulation is to address. In simulations that provide realistic representation of physical processes, the laws of physics and principles of engineering are part of the simulation context. For many military-related simulations, the simulation context includes standard organizational structures and general doctrine, strategy, and tactics. This is especially characteristic of C4ISR simulations. Often the simulation context is merely a collection of pointers and references to sources that define

behaviors and processes for things that will be represented within the simulation. Special care, especially for distributed interactive simulations[3], must be used when algorithms are taken from more than one source to ensure that sources do not employ contradictory assumptions or factors (such as different models for the shape of the Earth, differences in characterizing the environment, etc.). The information contained in the simulation context establishes boundaries on how the simulation developer can properly build the simulation.

> Example: This part of the conceptual model may identify such things as sources for the algorithms used for calculating radar signal propagation, operational modes possible with particular pieces of equipment, organizational structure and possible information-flow paths of a military unit, etc.

2. The **simulation concept** describes the simulation developer's concept for the entire simulation application (all the federates[4] and other pieces in a distributed simulation, i.e., everything that comprises the simulation) and explains how the simulation developer expects to build a simulation that can fully satisfy user-defined requirements. The simulation context establishes constraints and boundary conditions for the simulation concept. If the simulation is concerned with realistic representation of missile flight, then laws of physics and principles of aerodynamics are part of the simulation context, making the simulation concept accommodate conservation of momentum, etc. Unrealistic, cartoon representation of missile flight would not necessarily be so constrained. The simulation concept includes simulation elements, i.e., the things represented in the simulation. The simulation concept includes all simulation elements and specifies how they interact with one another: the simulation's mission space.

> Example: If the simulation is concerned with realistic representation of missiles or aircraft in flight, then the laws of physics and the principles of aerodynamics are part of the simulation context and require (constrain) the simulation concept to accommodate conservation of momentum, etc. Unrealistic, cartoon representations of missiles or aircraft in flight would not necessarily be so constrained.

The simulation concept has two primary aspects: mission space and simulation space:

- **Mission space** is concerned with representation. It includes the simulation elements (i.e., things and processes represented in the simulation).
- **Simulation space** is concerned with simulation control. It includes functional aspects of the simulation (e.g., hardware and software operating systems).

Thus, the simulation concept addresses both the representational aspects of the simulation (i.e., its mission space) and the functional aspects of the simulation (i.e., its simulation space). It specifies how they interact with one another and includes the additional information needed to explain how the simulation will satisfy the M&S requirements of the intended application.

The simulation space part of the simulation concept includes all additional information needed to explain how the simulation will satisfy its objectives. Such additional information often addresses control capabilities intended for the simulation, such as pause and restart capabilities, data collection and display capabilities, and how data and simulation control factors can be entered into the simulation (by keyboard, by voice, by gesture or touch, or by feedback from parts of the simulation). Simulation space characteristics range from identification of specific kinds of computing systems (hardware and operating systems) and timing constraints so that real systems can be part of the simulation (such as hardware in the loop unitary simulations or involvement of live forces in distributed simulations) to the kinds of simulation control capabilities as described above. Some simulation space considerations are closely related to implementation issues for the simulation. For example, selection of a parallel computing architecture has implications for algorithms used to describe simulation elements.

3. A **simulation element** consists of the information describing concepts for an entity, a composite or collection of entities, or process, which is represented within a simulation. It includes assumptions, algorithms, characteristics, relationships (especially interactions with other things within the simulation), data, etc. that identify and describe that item's possible states, tasks, events, behavior and performance, parameters and attributes, etc. A simulation element can address a complete system (such as a missile or radar), a subsystem (such as the antenna of a radar), an element within a subsystem (such as a circuit within a radar transmitter), or even a fundamental item (such as an atom). It can also address composites of systems (such as a ship with its collection of sensors, weapons, etc.). We should note that a person, part of a person (such as a hand), or a group of people can likewise be addressed by a simulation element. It can also address a process such as environmental effects on sensor performance.

> Examples:
> - A simulation element can address a complete system (e.g., a missile or radar), a subsystem (e.g., the antenna of a radar), an element within a subsystem (e.g., a circuit within the transmitter of a radar), or even a fundamental item of physics (e.g., an atom).
> - A simulation element can address composites of systems, such as a ship or aircraft with its collection of sensors and weapons, a person, part of a person (e.g., a hand), or a group of people.
> - A simulation element can address a process such as environmental effects on sensor performance.

Typical components of a simulation element are listed in the following table (Table 1-1).

Table 1-1. Example Components of a Simulation Element
• entity, process, or collection definition
• assumptions about, limitations of, and constraints placed on the element
• algorithms and algorithm pedigrees
• data and data history
• relations with other things within the simulation
• interactions with other things within the simulation

What information should a simulation conceptual model include?

A list of the types of information that should be considered for inclusion in a simulation conceptual model is provided in the following Table 1-2.

Table 1-2. Example List of Information Included in a Simulation Conceptual Model
1) Simulation descriptive information
• model identification (e.g., version and date)
• POCs
• model change history
2) Simulation context (per intended application)
• purpose and intended use statements
• pointer to M&S requirements documentation
• overview of intended application
• pointer to FDMS and/or other sources of domain information
• constraints, limitations, assumptions

• pointer to referent and referent information
3) Simulation concept (per intended application)
• mission space representation
o simulation elements
o simulation development environment descriptions (such as Unified Modeling Language (UML) entity/interaction diagrams)
• simulation space functionality
4) Simulation elements, including
• entity definitions (entity description, states, behaviors, interactions, events, factors, assumptions, constraints, etc.)
• process definitions (process description, parameters, algorithms, data needs, assumptions, constraints, etc.)
5) Validation history, including
• M&S requirements and objectives addressed in V&V effort(s)
• pointer to validation report(s)
• pointer to simulation conceptual model assessment(s)
6) Summary
• existing conceptual model limitations (for intended application)
• list of existing conceptual model capabilities (for intended application)
• conceptual model development plans

Notes

[1] Stakeholders are anyone who has an interest in the project. Simulation project stakeholders are individuals and organizations that are actively involved in the simulation project, or whose interests may be affected as a result of project execution or project completion. They may also exert influence over the simulation project's objectives and outcomes. The project management team must identify the stakeholders, determine their requirements and expectations, and, to the extent possible, manage their

influence in relation to the requirements to ensure a successful simulation project.

[2] The system lifecycle in systems engineering is an examination of a system or proposed system that addresses all phases of its existence to include system conception, design and development, production and/or construction, distribution, operation, maintenance and support, retirement, phase-out and disposal

[3] Distributed Interactive Simulation (DIS) is an open standard for conducting real-time platform-level wargaming across multiple host computers and is used worldwide, especially by military organizations but also by other agencies such as those involved in space exploration and medicine.

[4] A Federation is network of multiple simulations/simulators agreeing upon standards of operation in a collective fashion. Each simulation/simulator in the federation is called a federate.

Chapter 2. Simulation Conceptual Model Development

Development Process

A simulation conceptual model developed during simulation development provides a way of translating the M&S requirements of the intended application into a detailed design framework, from which the software, hardware, systems, and/or people that will make up the simulation can be built. There are five basic steps involved in developing a simulation conceptual model, which we can iterated a number of times throughout the development process as requirements change or modifications are made to design, data, or code. We list these steps below and discuss them in the following paragraphs.

1. Collect Authoritative Information
2. Decompose the Mission Space
3. Describe Simulation Elements
4. Identify Relationships
5. Assess and Record

1. Collect Authoritative Information.

Authoritative information is needed about the intended application domain that will constitute the simulation context, an important aspect of which is specification of the referent for fidelity and validity assessments (discussed at the end of this section). Collection of such authoritative information may involve the use of knowledge engineering techniques and knowledge acquisition/elicitation/representation processes developed for articulation of rules for expert systems [Knowledge Engineering website]; methods developed for problem formulation in operations research and systems analysis [5]; or, other formalisms employed in creating authoritative descriptions of entities, processes, and situations. However, development of the simulation concept and collection of authoritative information for the simulation context are likely to occur iteratively as the entities and processes to be represented become more clearly defined, regardless of which collection approach is taken.

> The Department of Defense Modeling and Simulation Coordination Office (MSCO), formerly the DMSO, developed descriptions of various military activities as part of the Functional Description of the Mission Space (FDMS)[1]. The FDMS descriptions (*"the first abstraction of the real world ... an authoritative knowledge source for simulation development ... capturing the basic information about important entities involved in any mission and their key actions and interactions"* [6]) can help to ensure commonality of perspective among various Defense simulations. Furthermore, they should facilitate reuse of simulation components, which will make both new simulation development and existing simulation modification more economical. Descriptions of military activities can be used for part of the simulation context when appropriate for a simulation's intended application, just as can the laws of physics and similar principles be used for other parts of the simulation context.

The formal, documented simulation context is unlikely to address everything needed to describe fully the domain that a simulation is to address. The FDMS endeavors described by Sheehan et al [7] emphasize a disciplined procedure that systematically informs the Developer about the real world and about a set of information standards that simulation SMEs employ to communicate with and obtain feedback from military operations SMEs. The keys to removing potential ambiguity between the ideas of the warfighting SMEs and the simulation development SMEs are

- common semantics and syntax
- common format database management system (DBMS)
- data interchange formats (DIF)

Significant progress has been made in developing a FDMS toolset to provide the keys noted above. However, their implementation to date, such as reported by Johnson [8], has shown that information may be required for simulation conceptual models beyond what is likely to be obtained in the first level abstraction (i.e., FDMS). Furthermore, SMEs may be "*called upon to fill in details needed by Developers*" that are "*not provided in doctrinal and/or authoritative sources*" [8]. Clearly, the more completely and clearly stated a simulation context is, the easier it will be to understand where and how one simulation may differ from another in its assumptions about the domain involved. This becomes very important when questions of compatibility among simulations (federates[2]) considered for a distributed simulation implementation (federation[3]) are addressed.

Sometimes it is obvious that we require additional information about the simulation context, if the simulation is to achieve its objectives (e.g., available information is inadequate, not just that it is not part of the authoritative description of the application domain). This often occurs for simulations used to support new system designs, and we establish test programs to generate such information. Sometimes the missing information consists only of parameter information (e.g., the strength of a Conceptual Model material or the signal level at which specified levels of distortion occur); other times, the missing information concerns the theory (or algorithms) used to describe entity behavior or performance.

> Example:
>
> How does the volume of a material change with temperature? Where does it change from solid to liquid to gas?

When significant information about critical aspects of a simulation is unknown or uncertain, development of the simulation conceptual model can be more difficult because the set of algorithms and data will be incomplete. The text by Patrick Roache [9] provides an excellent discussion of concerns about experimental (test) data, its limitations and uncertainties, its generation, and its relationship to simulation verification and validation (V&V). Sometimes we give inadequate attention to potential problems with the quality (correctness and comprehensiveness) of information upon which the simulation conceptual model is based.

2. Decompose the Mission Space

Simulation elements result from decomposition of the mission space, which defines the level of granularity or aggregation of the simulation. Six basic principles guide this decomposition:

Principles for Mission Space Decomposition
1. There should be a specific simulation element (parameter, entity, etc.) for every item (parameter, entity, etc.) specified for representation in the simulation-by-simulation requirements.
2. There should be a specific simulation element (parameter, entity, etc.) for every item (entity, task, parameter, state, etc.) of potential assessment interest related to the purpose of the simulation.
3. There should be "real world" counterparts (objects, parameters for which data exist or could exist, etc.) for every simulation element as far as possible. The potential impact of data, and metadata structures, on simulation elements and the simulation conceptual model should not be underestimated.
4. Wherever possible, the simulation elements should correspond to "standard" and widely accepted decomposition paradigms to facilitate acceptance of the conceptual model and effective interaction with other simulation endeavors (including reuse of algorithms and other simulation components).

5. Simulation elements required for computational considerations (e.g., an approximation used as a surrogate for a more desirable parameter that is not computationally viable) that fail to meet any of the previously stated criteria should be used only when essential.

6. There should not be extraneous simulation elements. Elements neither directly related to specific items in the simulation requirements nor directly implied by potential assessment issues and elements without a specific counterpart in the real world or in standard decomposition paradigms should not be included in the simulation conceptual model. Every extraneous simulation element is an unnecessary source of potential simulation problems.

To accomplish the application's objectives, the entities and processes that must be represented in the simulation should be identified by the six decomposition principles just listed. During this enumeration process, basic decisions are made about what level of detail and aggregation are appropriate to address the simulation requirements. These decisions determine whether a system (e.g., a radar) will be represented as a single entity, as a composite of subsystem entities (e.g., antenna, transmitter, receiver, etc.), or as a composite of composites of ever smaller entities to whatever level of detail is needed for the purpose of the simulation. Decisions are also made about the level of representation of human decisions and behaviors.

Figure 2-1 through 2-7 show possible conceptual modeling constructs for simulating a radar. One should clearly see that we made many assumptions—about requirements and other things—but do not explicitly state them. Thus, depending on your assumption, this is only one possible construct.

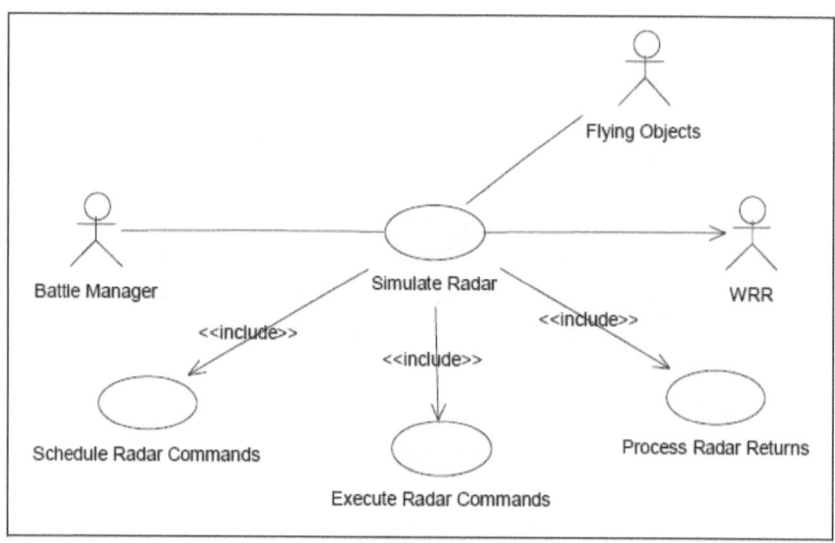

Figure 2-1. Top Level Sensor Use Case Diagram

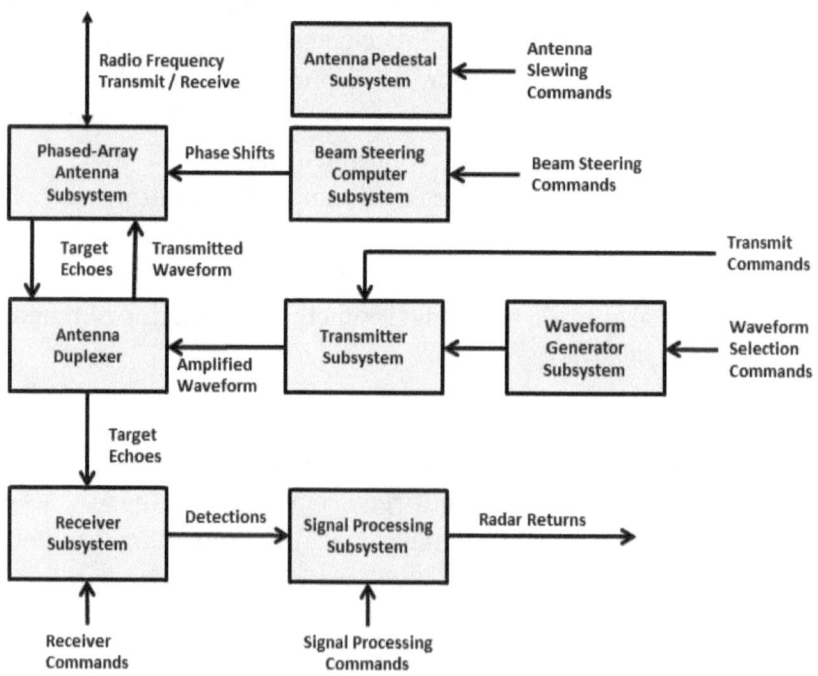

Figure 2-2. Break Down of Phased Array Radar

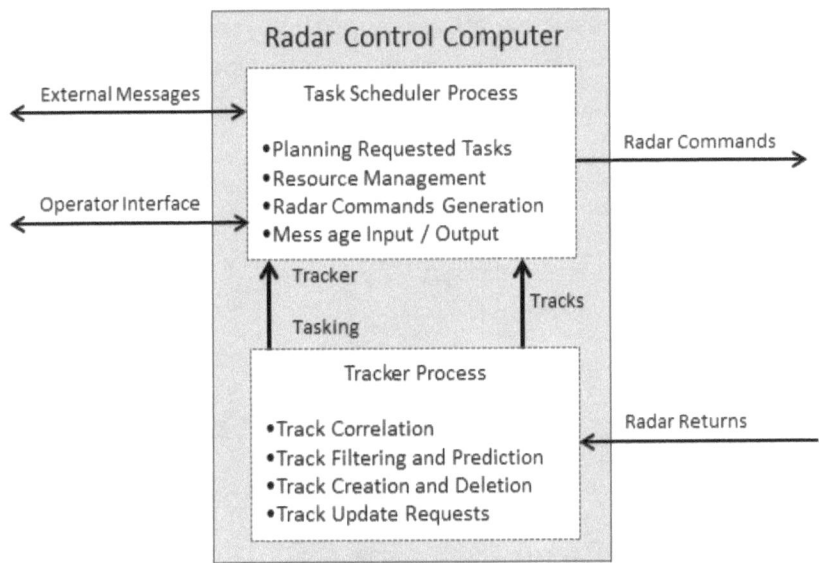

Figure 2-3. Radar Control Computer

> Example:
> In the operation of a radar (Sensor, Tracker, Controller, etc.), are the decisions and responses of all the subsystems/components involved (the sensor tasks) represented implicitly as a single aspect of the tracking process, or is each component involved represented explicitly (as in a tank simulator with a position for every member of the tank crew)?

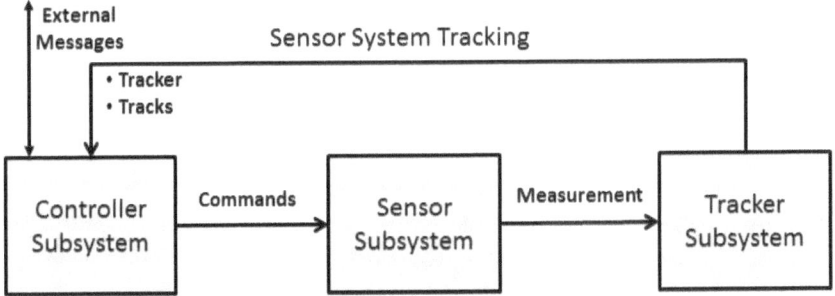

Figure 2-4. Generic Sensor System Conceptual View

Unfortunately, these theoretical approaches do not yet allow abstraction to be performed as a scientific method; abstraction remains an art [1].

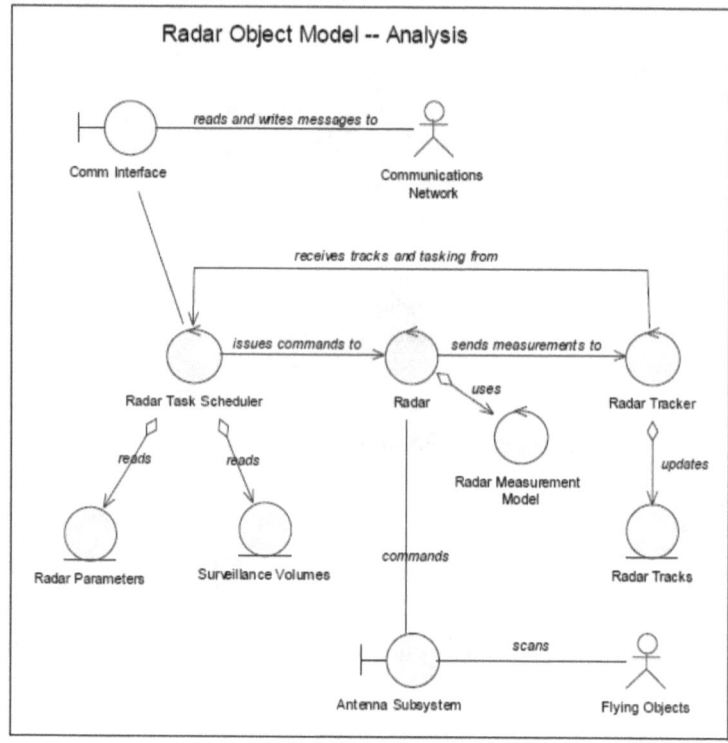

Figure 2-5. Radar Object Model

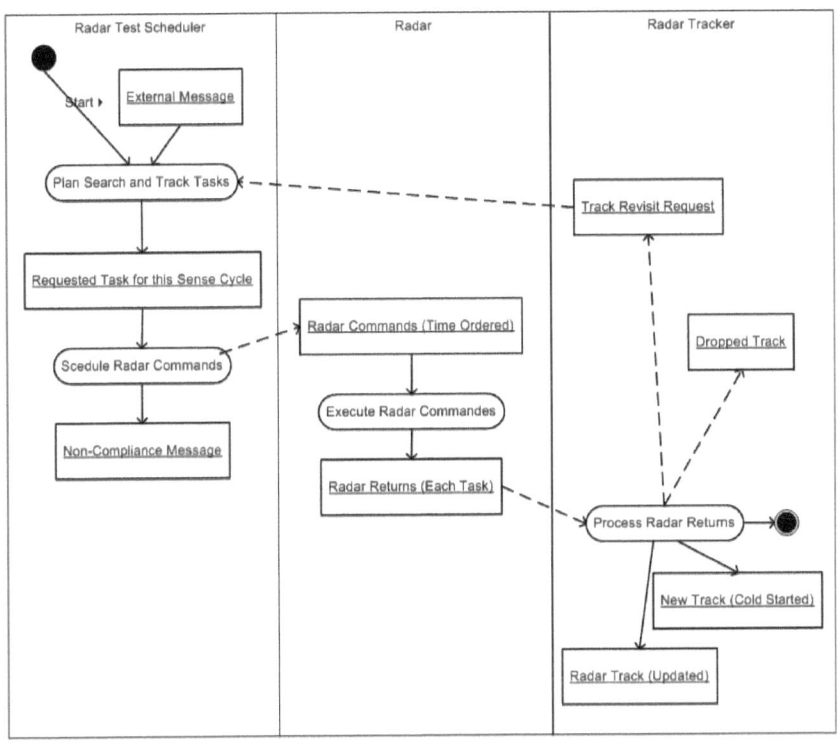

Figure 2-6. Radar Sense Cycle

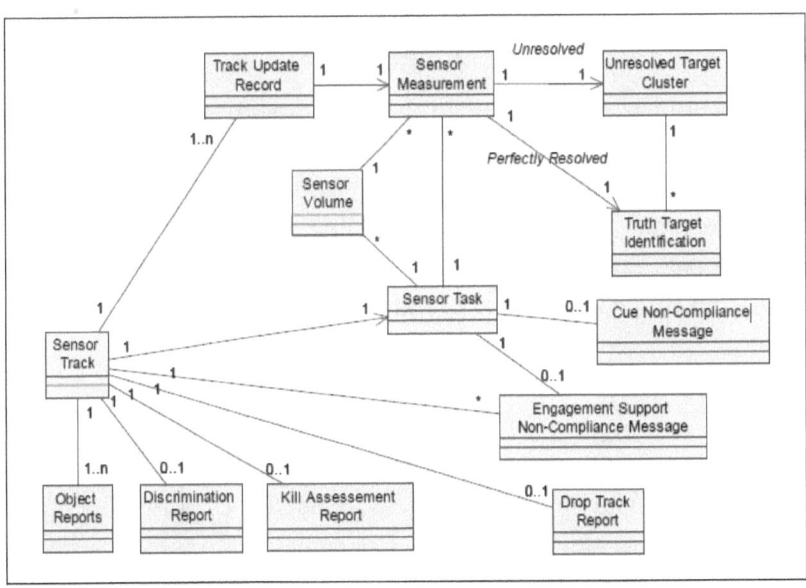

Figure 2-7. Notional view of a sensor data model

A review of recent articles in such publications as the *Journal of Data and Knowledge Engineering* [10] reveals that contemporary researchers in this arena often develop a "new" descriptive language (or dialect) or formalism for the problem at hand because current techniques do not yet have broad, general application capabilities.

As one develops a simulation conceptual model and evaluates it by the criteria of clarity, completeness, consistency, and correctness, it is important to record how one assesses the model and then to note why it changes in response to the evaluation, and how criteria for a quality conceptual model are met more fully. Otherwise, the rationale for some changes (and their benefits) may be lost as time passes, and lessons learned from the conceptual model development will not be readily available for use in subsequent developments.

The bottom line is simple: Consistent and comprehensive use of any formalism in conceptual model development is better than the common, ad hoc, unstructured approach frequently used.

4. Identify Relationships

The final step is to identify all of the relationships among simulation elements. This step should ensure that the constraints and boundary conditions imposed by the simulation context, as well as the operational and functional capabilities expressed in simulation requirements, are accommodated, and it should ensure that the simulation concept is fully articulated.

5. Assess and Record

As the simulation conceptual model is developed, we should evaluate it for clarity, completeness, consistency, and correctness as described in the section on Conceptual Model Assessment. The criteria used to define the level of quality needed, methods used in the assessment, and the results (e.g., which criteria are met and which are not) should be recorded as well as any changes made (to the conceptual model, M&S requirements, model design, etc.) based on the results of the assessment. The rationale for changes (and their benefits) and lessons learned from the conceptual model development can provide valuable information for subsequent use.

Development Considerations
Reality Abstraction

One should develop a simulation conceptual model within the larger context of simulation theory. The approach to abstracting reality into simulation terms is a key aspect of simulation theory. Without a coherent approach to such abstraction of reality, the different parts of the simulation are likely to be incompatible in some way with one another. A number of approaches to simulation theory are available. These include the approaches espoused by proponents of Application Domain Modeling (ADM) such as Hone and Moulding [11] and by proponents of the Discrete Event System Simulation (DEVS) methodology developed by Zeigler, et al [12]. Others prefer approaches to abstraction of reality for simulation based upon Casti's ideas [13]. Many within NASA stress use of formal methods where appropriate. The larger context of simulation theory can help ensure that simulation conceptual model development has coherence and can be related more directly to all aspects of simulation development.

Problem Identification

Development of a simulation conceptual model will often reveal problems with requirements for the simulation, especially if simulation requirements were not rigorously validated before the start of conceptual model development. As we develop the simulation conceptual model to fully satisfy simulation requirements, inconsistencies among requirements and lack of balance among the requirements (e.g., some very lax and others very stringent in the same general area) may become apparent. Development of the simulation conceptual model may also reveal serious holes in the requirements, areas that leave the Developer to his/her own initiative about what the simulation should be able to do. A well-structured simulation development program will encourage (if not insist upon) early formal and rigorous validation of simulation requirements and will ensure that requirement deficiencies uncovered during conceptual model development are corrected with appropriate modification to the simulation requirements.

Notes

[1] FDMS was previously known as the Conceptual Model of the Mission Space (CMMS).

[2] A Federate is a High Level Architecture (HLA) compliant simulation entity.

[3] A Federation is multiple simulation entities connected via the HLA Run-Time Infrastructure (RTI) using a common Object Model Template (OMT).

Chapter 3. Conceptual Modeling Techniques Overview

As systems have become increasingly complex, the role of conceptual modeling has dramatically expanded. With that expanded presence, the effectiveness of conceptual modeling at capturing the fundamentals of a system is being realized. Building on that realization, numerous conceptual modeling techniques have been created. These techniques can be applied across multiple disciplines to increase the users understanding of the system to be modeled [14]. We briefly describe a few techniques in the following text; however, many more exist or are being developed. Some commonly used conceptual modeling techniques and methods include; Data Flow Modeling, Entity Relationship Modeling, Event-Drive Process Chain, Joint Application Development, Place/Transition Net Modeling, State Transition Modeling, Object Role Modeling, and Unified Modeling Language (UML).

Structured System Analysis and Design Method

Structured Systems Analysis & Design Method (SSADM) is a widely used computer application development method in the UK, where its use is often specified as a requirement for government computing projects. It is increasingly being adopted by the public sector in Europe. SSADM is in the public domain, and is formally specified in British Standard BS7738[1].

SSADM divides an application development project into modules, stages, steps, and tasks, and provides a framework for describing projects in a fashion suited to managing the project (see Chapter 4).

Soft System Method

Soft Systems Methodology is an approach to inquiry into problem situations perceived to exist in the real world (Checkland and Scholes, 1990:18). It originates from the more general field of Systems Engineering, but have departed from the tradition of "hard" systems thinking (in which the perceived reality is considered systemic and inquiry systematic) into what is referred to as "soft" systems thinking (where perceived reality is problematic and inquiry is systemic). In their discussion of SSM and information systems development, [15] point out that SSM

> "is a framework which does not force or lead the systems analyst to a particular 'solution', rather to an understanding."

SSM has evolved through several versions, with [16] as the most widely cited (See Chapter 5).

Data Flow Modeling

Data flow modeling (DFM) is a basic conceptual modeling technique that graphically represents elements of a system. DFM is a simple technique; however, like many conceptual modeling techniques, it is possible to construct higher and lower level representative diagrams. The data flow diagram (see Chapter 8) usually does not convey complex system details such as parallel development considerations or timing information, but rather works to bring the major system functions into context. Data flow modeling is a central technique used in systems development that utilizes the Structured Systems and Analysis and Design Method (SSADM) (see Chapter 4).

Entity Relationship Modeling

Entity-relationship modeling (ERM) (see Chapter 10) is a conceptual modeling technique used primarily for software system representation. Entity-relationship diagrams, which are a product of executing the ERM technique, are normally used to represent database models and information systems. The main components of the diagram are the entities and relationships. The entities can represent independent functions, objects, or events. The relationships are responsible for relating the entities to one another. To form a system process, the relationships are combined with the entities and any attributes needed to describe further the process. Multiple diagramming conventions exist for this technique: IDEF1X, Bachman, and EXPRESS, to name a few. These conventions are just different ways of viewing and organizing the data to represent different system aspects.

Event-Driven Process Chain

The event-driven process chain (EPC) (see Chapter 11) is a conceptual modeling technique that is mainly used to improve systematically business process flows. Like most conceptual modeling techniques, the event driven process chain consists of entities/elements and functions that allow relationships to be developed and processed. More specifically, the EPC is made up of events that define what state a process is in or the rules by which it operates. In order to progress through events, a function/ active event must be executed. Depending on the process flow, the function has the ability to transform event states or link to other event driven process chains. Other elements exist within an EPC, all of which work together to define how and by what rules the system operates. The EPC technique can be applied to business practices such as resource planning, process improvement, and logistics.

Joint Application Development

The Dynamic Systems Development Method (DSDM) (see Chapter 13) uses a specific process called Joint Application Design (JAD) (see Chapter 12) to model conceptually a systems life cycle. JAD is intended to focus more on the higher-level development planning that precedes a projects initialization. The JAD process calls for a series of workshops in which the participants work to identify, define, and generally map a successful project from conception to completion. This method has been found not to work well for large-scale applications, however smaller applications usually report some net gain in efficiency [17].

Place/ Transition Net

Also known as Petri Nets (see Chapter 14), this conceptual modeling technique allows a system to be constructed with elements that can be described by direct mathematical means. The Petri net, because of its nondeterministic[2] execution properties and well-defined mathematical theory is a useful technique for modeling concurrent system behavior, i.e. simultaneous process executions.

State Transition Modeling

State transition modeling makes use of state transition diagrams (see Chapter 15) to describe system behavior. These state transition diagrams use distinct states to define system behavior and changes. Most current modeling tools contain some kind of ability to represent state transition modeling. The use of state transition models can be most easily recognized as logic state diagrams and directed graphs for finite state machines[3].

Object Role Modeling

Object Role Modeling (ORM) is a powerful method for designing and querying database models at the conceptual level, where the application is described in terms easily understood by non-technical users. In practice, ORM data models often capture more business rules, and are easier to validate and evolve than data models in other approaches [18].

Unified Modeling Language

The Unified Modeling Language™ (UML) is a standardized general-purpose modeling language in the field of object-oriented software engineering. Managed by the Object Modeling Group (OMG[4]), it is one of most-used specification. In addition, the way many system engineers model it can be used for not only application structure, behavior, and architecture, but also business process and data structure.

Notes

---

[1] British Standard 7738:1994 is a specification for information systems products using SSADM (Structured Systems Analysis and Design Method). Implementation of SSADM version 4

[2] In computer science, a nondeterministic algorithm is an algorithm that can exhibit different behaviors on different runs, as opposed to a deterministic algorithm. A probabilistic algorithm's behavior depends on a random number generator. An algorithm that solves a problem in nondeterministic polynomial time can run in polynomial time or exponential time depending on the choices it makes during execution.

[3] A finite-state machine (FSM) or finite-state automaton (plural: automata), or simply a state machine, is a behavioral model used to design computer programs. It is composed of a finite number of states associated to transitions. A transition is a set of actions that starts from one state and ends in another (or the same) state. A transition is started by a trigger, and a trigger can be an event or a condition.

[4] OMG is a not-for-profit computer industry specifications consortium

Chapter 4. Structured Systems Analysis and Design Method

Structured Systems Analysis and Design Method (SSADM) is a systems approach to the analysis and design of information systems. SSADM was produced for the Central Computer and Telecommunications Agency (now Office of Government Commerce), a UK government office concerned with the use of technology in government, from 1980 onwards.

Overview

SSADM is a waterfall method for the analysis and design of information systems. One can think of SSADM as representing a pinnacle of the rigorous document-led approach to system design, contrasting it with contemporary agile-methods—such as Dynamic Systems Development Method[1] (DSDM) or Scrum[2].

SSADM sets out a cascade or waterfall view of systems development, in which there are a series of steps, each of which leads to the next step. (We might contrast it with the rapid application development - RAD - method, which pre-supposes a need to conduct steps in parallel.). SSADM's steps, or stages, are [19]:

- Feasibility
- Investigation of the current environment
- Business systems options
- Definition of requirements
- Technical system options
- Logical design
- Physical design

For each stage, SSADM sets out a series of techniques and procedures, and conventions for recording and communicating information pertaining to these - both in textual and diagrammatic form. SSADM is a very comprehensive model, and a characteristic of the method is that projects may use only those elements of SSADM appropriate to the project. There are a number of CASE tools, which support SSADM.

SSADM is one particular implementation and builds on the work of different schools of structured analysis and development methods. For instance, Peter Checkland's Soft Systems Methodology (see Chapter 3), Larry Constantine's Structured Design (see Structured Analysis, Chapter 5), Edward Yourdon's Yourdon Structured Method, Michael A. Jackson's Jackson Structured Programming (see Structured Analysis, Chapter 7), and Tom DeMarco's Structured Analysis (see Chapter 5).

The names "Structured Systems Analysis and Design Method" and "SSADM" are registered trademarks of the Office of Government Commerce (OGC), which is an office of the United Kingdom's Treasury [20].

History

The principal stages of the development of SSADM were [21]:

- 1980: Central Computer and Telecommunications Agency (CCTA) evaluate analysis and design methods.
- 1981: Learmonth & Burchett Management Systems (LBMS) method chosen from shortlist of five.
- 1983: SSADM made mandatory for all new information system developments
- 1984: Version 2 of SSADM released
- 1986: Version 3 of SSADM released, adopted by NCC
- 1988: SSADM Certificate of Proficiency launched, SSADM promoted as 'open' standard
- 1989: Moves towards Euromethod[3], launch of CASE products certification scheme
- 1990: Version 4 launched

- 1993: SSADM V4 Standard and Tools Conformance Scheme Launched
- 1995: SSADM V4+ announced, V4.2 launched
- 1998: PLATINUM TECHNOLOGY acquires LBMS
- 2000: CCTA renamed SSADM as "Business System Development". The method was repackaged into 15 modules and another six modules were added [22] [23].

SSADM techniques

The three most important techniques that are used in SSADM are:

Logical Data Modeling
This is the process of identifying, modeling and documenting the data requirements of the system being designed. The data are separated into entities (things about which a business needs to record information) and relationships (the associations between the entities).

Data Flow Modeling
This is the process of identifying, modeling and documenting how data moves around an information system. Data Flow Modeling examines processes (activities that transform data from one form to another), data stores (the holding areas for data), external entities (what sends data into a system or receives data from a system), and data flows (routes by which data can flow).

Entity Behavior Modeling
This is the process of identifying, modeling and documenting the events that affect each entity and the sequence in which these events occur.

Stages

The SSADM method involves the application of a sequence of analysis, documentation and design tasks concerned with the following.

Stage 0 - Feasibility study

In order to determine whether a given project is feasible, there must be some form of investigation into the goals and implications of the project. For very small-scale projects, this may not be necessary, as we can easily understand the scope of the project. In larger projects, the feasibility may be done but in an informal sense, either because there is not time for a formal study or because the project is a "must-have" and will have to be done one way or the other.

When we carry out a feasibility study, there are four main areas of consideration:

- Technical - is the project technically possible?
- Financial - can the business afford to carry out the project?
- Organizational - will the new system be compatible with existing practices?
- Ethical - is the impact of the new system socially acceptable?

To answer these questions, the feasibility study is effectively a condensed version of a fully blown systems analysis and design. We analyze the requirements and users to some extent, draw up some business options and some details of the technical implementation.

The product of this stage is a formal feasibility study document. SSADM specifies the sections that the study should contain including any preliminary models that have been constructed and details of rejected options and the reasons for their rejection.

Stage 1 - Investigation of the current environment

This is one of the most important stages of SSADM. The developers of SSADM understood that though the tasks and objectives of a new system may be radically different from the old system, the underlying data would probably change very little. By coming to a full understanding of the data requirements at an early stage, the remaining analysis and design stages can be built up on a firm foundation.

In almost all cases, there is some form of current system even if it is entirely composed of people and paper. Through a combination of interviewing employees, circulating questionnaires, observations and existing documentation, the analyst comes to full understanding of the system as it is at the start of the project. This serves many purposes:

- the analyst learns the terminology of the business, what users do and how they do it
- the old system provides the core requirements for the new system
- faults, errors and areas of inefficiency are highlighted and their correction added to the requirements
- the data model can be constructed
- the users become involved and learn the techniques and models of the analyst
- the boundaries of the system can be defined

The products of this stage are:

- Users Catalog describing all the users of the system and how they interact with it
- Requirements Catalogs detailing all the requirements of the new system
- Current Services Description further composed of
- Current environment logical data structure (ERD)
- Context diagram (DFD)
- Leveled set of DFDs for current logical system
- Full data dictionary including relationship between data stores and entities

To produce the models, the analyst works through the construction of the models as we have described. However, the first set of data-flow diagrams (DFDs) are the current physical model, that is, with full details of how the old system is implemented. The final version is the current logical model, which is essentially the same as the current physical but with all reference to implementation removed together with any redundancies such as repetition of process or data.

In the process of preparing the models, the analyst will discover the information that makes up the users and requirements catalogs.

Stage 2 - Business system options

Having investigated the current system, the analyst must decide on the overall design of the new system. To do this, he or she, using the outputs of the previous stage, develops a set of business system options. One could produce the new system varying operation from doing nothing to throwing out the old system entirely and building an entirely new one in these different ways. The analyst may hold a brainstorming session so that as many and various ideas as possible are generated.

The analysts collect ideas to form a set of two or three different options that they present to the user. The options consider the following:

- the degree of automation
- the boundary between the system and the users
- the distribution of the system, for example, is it centralized to one office or spread out across several
- cost/benefit
- impact of the new system

Where necessary, the analysts document the options with a logical data structure and a level-1 data-flow diagram.

The users and analyst together choose a single business option. This may be one of the ones already defined or may be a synthesis of different aspects of the existing options. The output of this stage is the single selected business option together with all the outputs of stage 1.

Stage 3 - Requirements specification

This is probably the most complex stage in SSADM. Using the requirements developed in stage 1 and working within the framework of the selected business option, the analyst must develop a full logical specification of what the new system must do. The specification must be free from error, ambiguity and inconsistency. By logical, we mean that the specification does not say how the system will be implemented but rather describes what the system will do.

To produce the logical specification, the analyst builds the required logical models for both the data-flow diagrams[4] (DFDs) and the entity relationship diagrams[5] (ERDs). These are used to produce function definitions of every function, which the users will require of the system, entity life-histories (ELHs) and effect correspondence diagrams; these are models of how each event interacts with the system, a complement to entity life-histories. The analysts continually match these against the requirements and where necessary, the requirements are added to and completed.

The product of this stage is a complete Requirements Specification document, which is made up of the following:

- the updated Data Catalog
- the updated Requirements Catalog
- the Processing Specification which in turn is made up of
- user role/function matrix
- function definitions
- required logical data model
- entity life-histories
- effect correspondence diagrams

Though some of these items may be unfamiliar to you, it is beyond the scope of this unit to go into them in detail.

Stage 4 - Technical system options

This stage is the first towards a physical implementation of the new system. Like the Business System Options, in this stage a large number of options for the implementation of the new system are generated. The analysts hone this down to two or three to present to the user, from which the final option is chosen or synthesized.

However, the considerations are quite different being:

- the hardware architectures
- the software to use
- the cost of the implementation
- the staffing required
- the physical limitations such as a space occupied by the system
- the distribution including any networks which that may require
- the overall format of the human computer interface

All of these aspects must also conform to any constraints imposed by the business such as available money and standardization of hardware and software.

The output of this stage is a chosen technical system option.

Stage 5 - Logical design

Though the previous level specifies details of the implementation, the outputs of this stage are implementation-independent and concentrate on the requirements for the human computer interface. The logical design specifies the main methods of interaction in terms of menu structures and command structures.

One area of activity is the definition of the user dialogues. These are the main interfaces with which the users will interact with the system. Other activities are concerned with analyzing both the effects of events in updating the system and the need to make inquiries about the data on the system. Both of this use the events, function descriptions and effect correspondence diagrams produced in stage 3 to determine precisely how to update and read data in a consistent and secure way.

The product of this stage is the logical design, which is made up of:

- Data Catalog
- Required logical data structure
- Logical process model -- includes dialogues and model for the update and inquiry processes
- Stress and Bending moment.

Stage 6 - Physical design

This is the final stage where all the logical specifications of the system are converted to descriptions of the system in terms of real hardware and software. This is a very technical stage and we present a simple overview here.

The logical data structure is converted into a physical architecture in terms of database structures. We specify the exact structure of the functions and describe how we implement them. The physical data structure is optimized where necessary to meet size and performance requirements.

The product is a complete Physical Design, which could tell software engineers how to build the system in specific details of hardware and software and to the appropriate standards.

Advantages and disadvantages

Using this methodology involves a significant undertaking, which may not be suitable to all projects.

The main advantages of SSADM are:

- Three different views of the system
- Mature
- Separation of logical and physical aspects of the system
- Well-defined techniques and documentation
- User involvement

The size of SSADM is a big hindrance to using it in all circumstances. There is a large investment in cost and time in training people to use the techniques. The learning curve is considerable as not only are there several modeling techniques to come to terms with, but there are also a lot of standards for the preparation and presentation of documents.

Notes

[1] Dynamic systems development method (DSDM) is an agile project delivery framework, primarily used as a software development method. DSDM was originally based upon the rapid application development method. In 2007, DSDM became a generic approach to project management and solution delivery. DSDM is an iterative and incremental approach that embraces principles of Agile development, including continuous user/customer involvement.

[2] Scrum is an iterative, incremental framework for project management often seen in agile software development, a type of software engineering.

[3] Euromethod was a method for managing procurement processes of Information Services. It focuses on contract management. Euromethod consisted of three books: a reference manual, a dictionary and a collection of annexes. Euromethod's first release was in 1996. Although Euromethod is a technically sound method it was never widely known or used.

⁴ A data flow diagram (DFD) is a graphical representation of the "flow" of data through an information system, modeling its process aspects. Often they are a preliminary step used to create an overview of the system, which can later be elaborated.

⁵ In software engineering, an entity-relationship model (ERM) is an abstract and conceptual representation of data. Entity-relationship modeling is a database modeling method, used to produce a type of conceptual schema or semantic data model of a system, often a relational database, and its requirements in a top-down fashion. Diagrams created by this process are called entity-relationship diagrams, ER diagrams, or ERDs.

Chapter 5. Soft systems methodology

Soft systems methodology (SSM) is a systemic approach for tackling real-world problematic situations [24]. Soft Systems Methodology is the result of the continuing action research that Peter Checkland[1] [25], Brian Wilson[2] [26], and many others [27] have conducted over 30 years, to provide a framework for users to deal with the kind of messy problem situations that lack a formal problem definition [28] [29].

Overview

A common misunderstanding is that SSM is a methodology for dealing solely with 'soft problems' (i.e., problems which involve psychological, social, and cultural elements). SSM does not differentiate between 'soft' and 'hard' problems; it merely provides a different way of dealing with situations perceived as problematic. The 'hardness' or 'softness' is not the intrinsic quality of the problem situation to be addressed, it is an aspect of the way those involved address the situation. Each situation perceived as problematic has both 'hard' and 'soft' elements. The very notion of a problem is contingent on a human being perceiving it as such, namely, "One man's terrorist is another man's freedom fighter" [30].

SSM distinguishes itself from hard systems approaches in the way it deals with the notion of 'system'. Common to hard systems approaches is an understanding of systems as ontological entities, i.e., entities existing in the real world. As such, in hard systems approaches when one speaks of a computer system, an information system, a telecommunications system, or a transport system, one refers to these as bounded entities with a physical existence which can be formally described or designed to fulfill a given purpose.

In contrast, SSM treats the notion of system as an epistemological[3] rather than ontological[4] entity, i.e., as a mental construct used for human understanding. If we look for example at a particular organization as a system, we can describe this organization as a system to make a profit, or a system to transform raw materials into a commercial product, or a system to provide jobs to the local community, or a system to pollute the environment. Depending on what perspective we take, we will have a very different understanding of this particular organization.

None of these descriptions is right or wrong, they are merely different ways of understanding what is going on. This requires us to become conscious of our particular perspective and values, and these in turn determine what aspects of the situation we understand as being part of the system of our concern. For instance, if we are trying to understand this organization as a system to transform raw materials into a commercial product, we are likely to include the providers of raw materials and the customers who buy the end-product in our understanding of this system. However, if we look at the organization as a system to provide jobs to the local community, we are likely to include different elements such as the local transport infrastructure, which allows members of the community to access the organization. As such, depending on our perspective we draw different boundaries around what we perceive the system to be.

History

Developed from 1966 by a team of academics from the University of Lancaster, led by Prof Gwilym Jenkins, and resulted from their attempts to tackle management problem situations using a systems engineering approach. The team found that Systems Engineering, which was a methodology so far only used for dealing with technical problems, proved very difficult to apply in real-world management problem situations. This was especially so because the approach assumed the existence of a formal problem definition. However, it was found that such a unitary definition of what constitutes 'the problem' was often missing in organizational problem situations, where different stakeholders often have very divergent views on what constitutes 'the problem'.

SSM has received its fame and recognition through the work of Prof Peter Checkland who joined the team in 1969 appointed as the new Professor of "Commercial" systems and Dr. Brian Wilson who had joined in 1966 and ran the action research program through the University's consulting arm 'ISCOL' from 1970.

The development of SSM lends itself particularly well to dealing with complex situations, where those involved lack a common agreement on what constitutes the problem, and that needs to be addressed. In such situations (namely, How to improve health service delivery; How to conduct a business in a more sustainable way; How best to deal with youth offenders; or How best to deal with drug abuse), there may exist many different perspectives, values, and beliefs around what aspects of the situation are most important and how to address them. Those various aspects perceived as problematic tend to be highly interrelated; changing one aspect is likely to have knock-on effects on other aspects. It is important therefore to develop a comprehensive understanding of those interrelationships between the various aspects of the problem situation. As a systemic methodology, SSM aims to aid its users in developing an improved understanding though an iterative learning process.

As an offspring of Enid Mumford's 1960's "Participative Approach" (and sometime MSc. External Examiner at Lancaster), stakeholders are likely to reach accommodations – agreements about what changes to the situation the participating parties can live with. The notion of accommodation needs to be distinguished from the concept of consensus. Consensus implies that all the stakeholders fully agree that the proposed changes best serve all of their needs. The concept of accommodation recognizes that this is a very rare state of affairs in most real-world situations, and that most of the time individual needs can only be partially met by collective propositions.

The 7-Stage Approach of SSM

The original version of SSM as a seven-stage methodology published in Checkland's "Systems Thinking, Systems Practice" [25] has since been superseded in Checkland's work. However, the seven-stage model is still widely used and widely taught because its step-wise nature makes it easily teachable. Most important, the model has a barrier running across it to differentiate stages between the Real World, above the line, and Systems Thinking, below the line; the rigor to the method and a latter day *pons asinorum* for many students,

The seven stages are:

1. Entering the problem situation.
2. Expressing the problem situation.
3. Formulating root definitions of relevant systems.
4. Building Conceptual Models of Human Activity Systems.
5. Comparing the models with the real world.
6. Defining changes that are desirable and feasible.
7. Taking action to improve the real world situation.

The dynamics of the method come from the fact that stages (2) through (4) are always an iterative process. The stakeholders (defined as Client, Actors and Owner) engage in a debate guided by the analyst/facilitator. During this debate, various root definitions (succinct statements of appropriate systems) and conceptual models are put forward, modified and developed until a desirable model is achieved by consensus. This model then forms the basis for real world changes [31].

CATWOE

The Lancaster team proposed several criteria that should be specified to ensure that a given root definition is rigorous and comprehensive. These criteria are summarized in the mnemonic *CATWOE* [32]:

Clients – Who are the beneficiaries or victims of this particular system? (Who would benefit or suffer from its operations?)

Actors – Who are responsible for implementing this system? (Who would carry out the activities that make this system work?)

Transformation – What transformation does this system bring about? (What are the inputs and what transformation do they go through to become the outputs?)

Weltanschauung (or **Worldview**) – What particular worldview justifies the existence of this system? (What point of view makes this system meaningful?)

Owner – Who has the authority to abolish this system or change its measures of performance?

Environmental constraints – Which external constraints does this system take as a given?

This form of analysis clarifies what the user of the methodology is trying to achieve. By explicitly acknowledging these perspectives, the user of the methodology is forced to consider the impact of any proposed changes on the people involved.

Conceptual Models of Human Activity Systems

SSM Conceptual Models of Human Activity Systems (Conceptual Models) are notional, they are not intended to represent what exists but to represent a stakeholder viewpoint [26] [27]. This is often misunderstood. Figure 5-1 is not intended to represent how rice is cooked; but how the stakeholders 'think' it is cooked or how they 'think' it should be cooked or how they would like it cooked.

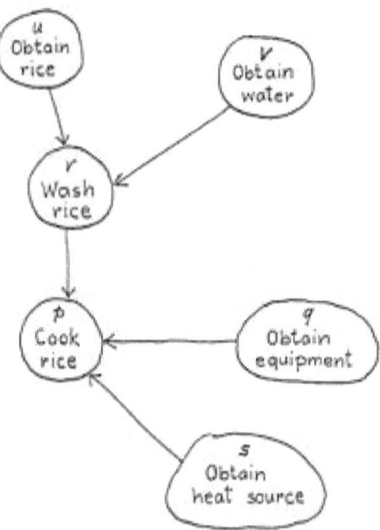

Figure 5-1. An SSM style Conceptual Model

Conceptual Models take the form of bubble diagrams in which descriptions of activities are enclosed in bubbles and the bubbles linked to each other by arrows. The arrows are intended to represent logical dependency. In Figure 5-1 the activity "wash rice" is said to be logically dependent on the activities "obtain rice" and "obtain water" being performed. This relation of "logical dependency" is transitive, i.e. if *cook rice* is dependent on *wash rice* and *wash rice* is dependent on *obtain rice*, then *cook rice* is dependent on *obtain rice*. This would appear to conform to what is known, in formal logic, as hypothetical syllogism[5]. However, a connection with logic has been challenged and it has been argued that SSM conceptual models are not "logical" in any sense of the word [33].

In Checkland's work [27] [34] Conceptual Models are usually limited to a small number (seven, plus or minus two) of bubbles. In addition, in fidelity to Cybernetics, the main activities are always supplemented by bubbles representing monitor and control systems. However, in Brian Wilson's "Information Requirements Analysis" [26], the Conceptual Models may expand to include hundreds of bubbles, and the monitor and control systems are dropped. While the principal SSM authors show a high degree of similarity in their accounts of the early stages of the method, considerable diversity begins to appear at the Conceptual Model building stage.

Outcomes and applications of SSM

General descriptions of SSM are highly diverse. SSM has been characterized as a learning system, part of a new paradigm for Operations Research[6] and as a front-end for information system design. However, such diversity is to be expected, considering that its aim is to address any kind of unstructured "soft" problem in any organizational or social context. SSM functions as a learning system because it facilitates a greater understanding of the problem situation on the part of those concerned. By bringing out the world views (Weltanschauung) of the people involved in the problem situation, SSM can produce various types of result. The problem might simply disappear as the result of a consensus. An unstructured solution might result, such as agreement to adopt a new role for the organization. A third possibility is that the problem becomes structured, in this case a soft problem resolves into an identifiable "hard" problem [31]. SSM has been used extensively in Information Systems Analysis and Design and some information systems textbooks treat SSM purely as a systems analysis and design method (see Curtis [35]).

The results of a survey of SSM in practice were published in 1992 [36]. Based on respondents' answers to an open-ended question, the following applications were identified:

Organizational design: Restructuring of roles, design of new organizations, and creation of new organizational culture.

Information systems: Definition of information needs, creating an IS strategy, knowledge acquisition, evaluation of the impact of computerization.

General problem solving: Understanding complex situations, initial problem clarification.

Performance evaluation: Performance indicators, quality assurance, monitoring an organization.

Education: Defining training needs, course design, causes of truancy, analysis of language teaching.

Miscellaneous: Project management, business strategy, risk management methodology, case for industrial tribunal, personal life decisions.

SSM for Information Systems Analysis and Design

The uses of SSM in information system design are many and varied. Some of the most notable methods are:

Checkland and Holwell
Checkland and Holwell [34] use SSM at the front end of an information systems design project. Their projects have been concerned with the reorganization of an information systems department, the evaluation of information systems and developing information systems strategy.

Their work does not extend into software engineering and is confined to analyzing the scope and facilitating the management of an information systems design project. As such, it has been comparatively free from criticism.

Information Requirements Analysis
Information Requirements Analysis [26] (IRA) seeks to identify the information required in a client organization by building conceptual models comprising hundreds of bubbles. We use these models to derive "information categories" and map activity to activity to activity information flows on a matrix known as a "Maltese Cross". IRA links directly to software design and has application in building transaction-processing systems.

Unlike the Checkland and Holwell projects, where the models remain at a conceptual level, IRA seeks to build models for the design of information systems that can provide information about real world objects and events (such as stock control systems). It has been argued that IRA models do not have the logical power to represent cause and effect and, therefore, an information system built out of them cannot represent events in the physical world [31].

The use of IRA has not been, however, limited to building transaction processing systems. For example, IRA was used in undertaking an audit of an analysis method — Microanalysis — for improving effectiveness and efficiency in a particular area of policing known as 'protective services'. The Regional Project Director was tasked with exploring options for collaboration between North Yorkshire Police, South Yorkshire Police, West Yorkshire Police and Humberside Police. SSM was used to develop a reference model relevant to protective services, which, together with information categories for each of the SSM activities and conceptual measures of performance, was used to analyze the efficacy of Microanalysis by comparing and contrasting information content [37].

Multiview
Multiview [38] seeks to front end SSM onto established software engineering methods such as SSADM and Information engineering. Multiview builds Conceptual Models and derives Data Flow Diagrams and Entity-relationship Models from them. Multiview links directly to software design and has application in building transaction-processing systems.

The Multiview Conceptual Models are not notional and appear to represent things in the physical world. While this obviates some of the theoretical problems found in IRA, it loses some of the advantages of traditional SSM and opens up a set of problems found in other information system design methods [39].

Logico-linguistic Modeling

Logico-linguistic modeling [40] uses logically enhanced Conceptual Models for Knowledge Elicitation and Representation. These models can be expressed in modal predicate logic from which code in the Prolog artificial intelligence language can be derived. Logico-linguistic Modeling has application in Knowledge based system design.

While Logico-linguistic Modeling overcomes the problems in the transition from conceptual model to computer code, it does so at the expense of making the stakeholder constructed models much more complex. It has been argued that the benefit of this complexity is questionable [41] and that this modeling method is much harder to use [42].

Notes

[1] Peter Checkland (1930 Birmingham, UK) is a British management scientist and emeritus professor of Systems at Lancaster University. He is the developer of soft systems methodology (SSM): a methodology based on a way of systems thinking.

[2] Brian Wilson (born 1933, in Newton-le-Willows, Lancashire) is a British systems scientist and honorary professor at Cardiff University, known for his development of soft systems methodology (SSM) and enterprise modeling.

[3] Epistemology (from Greek ἐπιστήμη (epistēmē), meaning "knowledge, science", and λόγος (logos), meaning "study of") is the branch of philosophy concerned with the nature and scope (limitations) of knowledge. It addresses the questions: What is knowledge? How is knowledge acquired? How do we know what we know? Much of the debate in this field has focused on analyzing the nature of knowledge and how it relates to connected notions such as truth, belief, and justification. It also deals with the means of production of knowledge, as well as skepticism about different knowledge claims.

[4] Ontology (from onto-, from the Greek ὤν, ὄντος « being; that which is », present participle of the verb εἰμί « be », and -λογία, -logia: science, study,

theory) is the philosophical study of the nature of being, existence or reality as such, as well as the basic categories of being and their relations. Traditionally listed as a part of the major branch of philosophy known as metaphysics, ontology deals with questions concerning what entities exist or can be said to exist, and how such entities can be grouped, related within a hierarchy, and subdivided according to similarities and differences.

[5] In logic, a hypothetical syllogism has two uses. In propositional logic, it expresses one of the rules of inference, while in the history of logic; it is a short-hand for the theory of consequence.

[6] Operations Research (also referred to as decision science, or management science) is an interdisciplinary mathematical science that focuses on the effective use of technology by organizations. In contrast, many other science & engineering disciplines focus on technology giving secondary considerations to its use. Employing techniques from other mathematical sciences — such as mathematical modeling, statistical analysis, and mathematical optimization — operations research arrives at optimal or near-optimal solutions to complex decision-making problems. Operations Research is often concerned with determining the maximum (of profit, performance, or yield) or minimum (of loss, risk, or cost) of some real-world objective. Originating in military efforts before World War II, its techniques have grown to concern problems in a variety of industries

Chapter 6. Soft Model Analysis for Discrete Event Simulation (DES)

Overview

Soft model differs from conventional (hard) modeling in that, unlike hard modeling, it is tolerant of imprecision, uncertainty, partial truth, and approximation. In effect, the role model for soft computing is the human mind. The guiding principle of soft modeling is this: Exploit the tolerance for imprecision, uncertainty, partial truth, and approximation to achieve tractability, robustness and low solution cost. The basic ideas underlying soft modeling in its current incarnation have links to many earlier influences, among them Zadeh's 1965 paper on fuzzy sets[1] [43]; the 1973 paper on the analysis of complex systems and decision processes [44]; and the 1979 report (1981 paper) on possibility theory and soft data analysis [45]. The inclusion of neural networks[2] and genetic algorithms in soft modeling came at a later point.

At this juncture, the principal constituents of Soft Modeling (SM) are Fuzzy Logic (FL), Neural Networks (NN), Evolutionary Algorithms (EA) Machine Learning (ML) and Probabilistic Reasoning (PR), with the latter subsuming Bayesian belief networks[3], chaos theory[4] and parts of learning theory[5]. What is important to note is that soft modeling is not a mélange. Rather, each of the partners contributes a distinct methodology for addressing problems in its domain in a partnership. In this perspective, the principal constituent methodologies in SM are complementary rather than competitive. Furthermore, soft modeling may be viewed as a foundation component for the emerging field of conceptual intelligence.

In system dynamic models introduced by Forrester [46], a soft approach is the construction of "causal loop diagrams"[6] [47]. The soft model represents the feedback behavior of dynamic systems. Contrary to dynamic systems, variables in a DES model rarely present feedback behavior. Thus, it is possible to create a "soft model" by making a direct mapping between input and output variables, once the discrete event conceptual model has been created.

> Example: Consider a single-server queue simulation model. What happens to the entity's waiting time in the queue (output) if the entity's time between arrivals is shorter (input)? Very often, we know the general behavior of several input-output relationships before the model runs. If the modeler does not know, the system's expert can infer if easily based on experience.

The purpose of soft modeling is to generate debate and gain more insight about the real world. The soft methodology can be implemented using Design of Experiments[7] (DOE). Soft modeling depends heavily upon subject matter expert (SME) input and the simulation conceptual model.

Analysis Preparation
(Step 1) Determine Prerequisites

The primary input to this process is the DES conceptual model documentation. Because the formality of the documentation for projects tends to vary widely, the required inputs may be in the form of formal notes, formal documentation, and magnetic media accessible through automated development tools or even non-existent. The term documentation is used to indicate any of these sources or others in which the necessary information exists:

- Conceptual model documentation
- Intended Use Statement
- System/software/interface specification documentation
- System design documentation

If there is no formal conceptual model, system design and specification documentation may be used to develop the soft model. If there is no Intended Use Statement available for the model or simulation, it must be developed before initiating this procedure.

The intended use of the model or simulation may be specified by a general statement of a set of potential uses or specified in a detailed test plan including an experiment design, scenario, and threat. The more detailed the intended use information, the better the results of this procedure. A list of the functional model capabilities required is developed by the users and model developer and if not available may be developed before performing this procedure. A specific description of the intended use will allow for a more precise assessment of the model.

(Step 2) Select Analyst Subject Matter Experts (SMEs)

If a team leader has not been identified yet, this selection should precede the identification of any SME. The team leader is needed to coordinate the procedure, document its results, and assist in SME selection. He or she should identify SME skills and assign team responsibilities (i.e., identify which analyst is to examine each sub model).

SME availability is critical. Analysts selected to perform this procedure should be experienced in general model or simulation utilization, in physics, in mathematics, or in software development practices and procedures. Most importantly, they should have a background in the functionality or sub-model for which they are responsible. Previous experience in the utilization and analysis of a "like type" model or simulation is also beneficial. For small models or simulations, a single analyst can perform this effort; for larger models, however, multiple sub models can be partitioned across a team.

As the SMEs are selected, distribute all available background materials, this procedure description, and information about schedule requirements (when the effort is to be completed and how much effort has been allocated).

(Step 3) Assemble Conceptual Model Documentation

This can be a challenge as conceptual models are often not formally constructed or documented. A simulation conceptual model consists of three categories of information about the simulation and the intended application(s)/uses(s), the simulation context, simulation concept, and simulation elements, shown in the Figure 6-1.

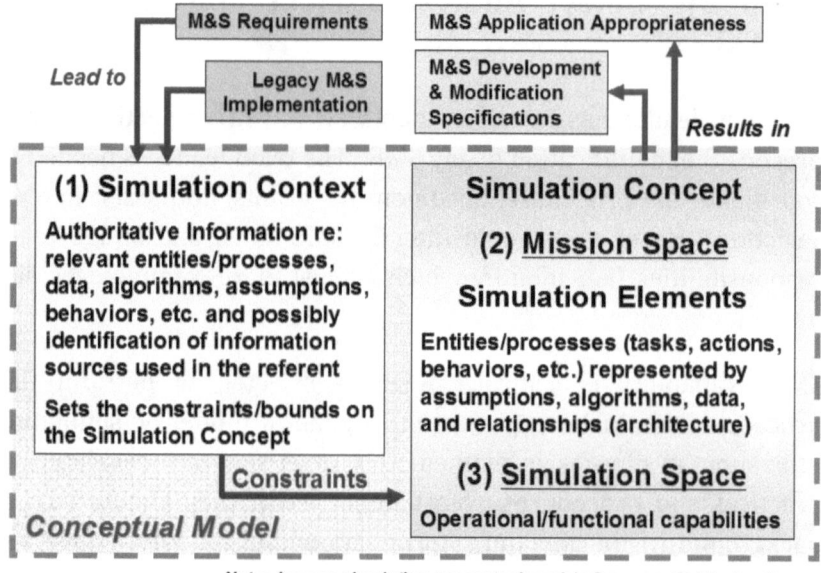

Figure 6-1. Conceptual Model Components

The simulation context sets the constraints on the simulation context. It should document the relevant entities and processes, as well as data, algorithms, assumptions, and behaviors. The simulation concept describes the mission space and the simulation space. It also documents the problem space when the simulation is designed to support problem solving, as in a studies and analysis, or analysis of alternatives[8] (AOA). The mission space includes the entities and processes that are represented of the assumptions, algorithms, data, and architecture of the proposed simulation, as depicted by the architecture. The simulation space includes the operational and functional capabilities represented by the proposed simulation.

(Step 4) Select Input and Output Variables

The analyst, with SME support, should choose variables that will result in a "feasible" causal influence matrix[9] and DOE. Choosing 100% of the model input and output variables are not feasible, particularly when modeling complex systems. The analysts should select input and output variables based on corresponding critical system design factors.

(Step 5) Develop a Soft Model

DOE is a general method for specifying a soft model of the simulation, based on SME input and conceptual model review. Another approach uses simple causal influence matrices. One soft model can be summarized as simply a causal influence matrix, C, we construct in the following manner: Given N input variables $(I_1, I_2, I_3, ..., I_N)$ and M output variables $((O_1, O_2, O_3, ..., O_M)$ of the simulation model, the component C_{NM} of the correlation matrix[10] can assume the vales of -1, 0, and 1 (or simply "−", "+", "0") which indicates respectively a negative, neutral, or positive correlation. As an example, Figure 6-2 shows a causal influence matrix of 3 inputs and 3 outputs.

	O_1	O_2
I_1	+1	-1
I_2	+1	0
I_3	-1	0

Figure 6-2. Causal influence matrix example

In this case, observe that the correlation of I_1 and O_1 is positive: this means that when I_1 rises, O_1 rises. However, the correlation of I_3 and O_1 is negative: when I_3 rises, O_1 decreases or vice-versa. I_2 and O_2 possess a neutral correlation: in this case the variation of I_2 will not affect O_2. In the single server queue example, the correlation between an entity's arrival and waiting time in the queue is negative (higher times between arrivals implies lower waiting times in the queue). It is fundamental that this matrix be built by an experienced simulation analyst and/or system expert, and not through simulation runs.

It is important to notice that the matrix is built upon two principles:

1. Linearity of input / output
2. No crossed correlation between input variables (interdependencies).

This is why the matrix is easy to build from scratch: human thinking is linear and not correlated. However, the variables in the model can express non-linear behavior and a correlation may exist between variables.

(Step 6) Build the Discrete Event Model

After developing the conceptual model and the soft model, the modeler build a computational model of the system, in this case a DES model. However, in most cases within DoD, we are using legacy models and/or iterating models for spiral builds. Consequently, the DES model is usually already built and we only require modifications.

(Step 7) Build the DOE and Make Simulation Runs

Next, the analysts builds the DOE (see Montgomery, 2008 [48]) and makes simulation runs in order to construct the correlation matrix of main effects.

(Step 8) Apply the Validation Technique

The proposed V&V technique for DES starts by building the causal influence matrix (soft model). This activity takes place after the definition of the conceptual model or the computerized model, since input and output variables must be chosen. After the implementation of the model, we must confront the values obtained from the matrix against the values in the results of the simulation runs. One relatively simple way to do this is implementing a 2^k factorial experimental design[11] (for a detailed description of this technique, refer to Montgomery 1984). Now let us suppose that we have done the 2^k factorial design with the same Input and Output of the causal influence matrix as depicted in Figure 6-2. If we take into account only the main effects, we could obtain the results shown in Figure 6-3.

	O_1	O_2
I_1	+40.8	- 12.9
I_2	+15.27	0.002
I_3	- 16.18	- 0.012

Figure 6-3. 2^k factorial experimental design

By looking at the DOE results, we know that a high positive value means a positive correlation, a high negative value means a negative correlation, and a value near 0 means that the given input has negligible effect over the output variable. Therefore, if we compare directly the 2^k factorial design with the causal influence matrix shown in Figure 4-2, we see no discrepancies. If this is true, the process is complete.

Now suppose that we have some kind of discrepancy. For instance, in the causal influence matrix, we have a positive correlation, but when we do the factorial design, it shows a null or negative correlation. Therefore, we have to look deeper for the source of this discrepancy. Discrepancies may occur due to three distinct factors (for brevity, we call soft model the causal influence matrix, and hard model the computerized model):

- The soft model is "correct" but there is some problem with the hard model.
- The hard model is "correct" but the behavior addressed by the soft model is erroneous.
- The soft model does not match the hard model because of inherent limitations of the soft model as mentioned before (non-linearity and correlated inputs).

The discrepancies generated by the factors 1 and 2 will cease if we either understand the behavior of the hard model and fix the soft model, or if we alter the computerized model to match the pattern predicted by the soft model. In this latter case, we will have to modify the model and redo the DOE. In the case of the third factor, the modeler should make plots to look for non-linearities and verify the values of the 2nd and higher order interactions that stem from 2^k analyses, to verify the interactions between variables. In the former case, if a non-linearity is detected, the modeler can correct the soft model to address the real correlation or change the factorial levels. For instance, let us suppose that for a theater missile defense with uniform assignment of targets, the overall attack assessment is carried out first before the defensive missiles (interceptors) are fired to minimize any unnecessary overlap [49]. Now, let W_i represent the number of warheads penetrating defense layer i. W is the total number of attacking warheads. $P(D_i)$ is the probability of destruction in layer i. If N is the number of interceptors, and N_1 is the number of interceptor in the first layer, then N/W is the ratio of total interceptor and attacking warheads, while N_1/W is the ratio of interceptors in layer one and total attacking warheads. Likewise, W_1/W is the ratio of attacking warheads penetrating layer one and the total attacking warheads. Theoretically, we can expect that as ratio N/W to increases, the ratio W1/W decreases, as depicted in

Figure 6-4. That is, as the total number of interceptors "outnumber" the total number of attacking warheads, the ratio of number of attacking warheads penetrating layer one to the total number of warheads decreases. Thus, the correlation between W1/W and N/W is negative.

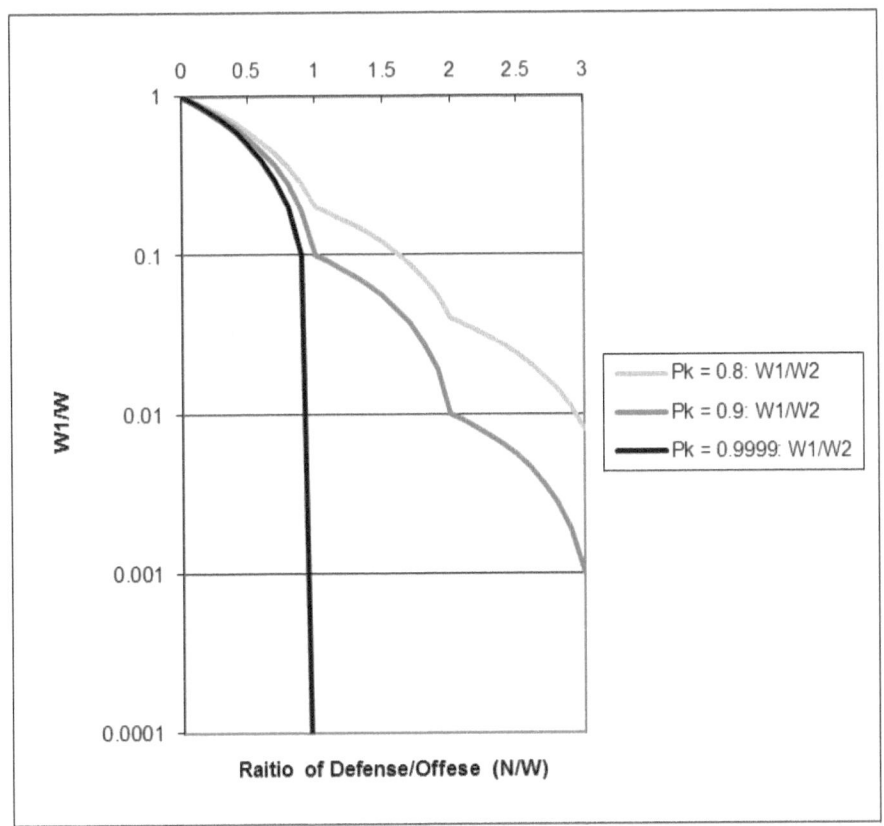

Figure 6-4. Ratio of Defense to Offense in Theater BMD

Now let us suppose the soft model shows no correlation between $W1/W$ and N/W (this only occurs when N/W is a positive integer (i.e., 0, 1, 2, ...). Therefore, we can either alter the soft model or alter the factorial levels to match the same range of input variable (e.g., (0, 1), (1, 2), and (2, 3) in Figure 6-4).

In the case of correlation of inputs (or interactions), if higher order interactions are not negligible (this is easily seen by 2^k

experiments), it is advisable to drop the variable that causes the interactions and consider others that are independent from each other. However, if the interaction is well understood, the modeler could gain valuable insight about the behavior of the computerized model.

We summarize this approach in Figure 6-5.

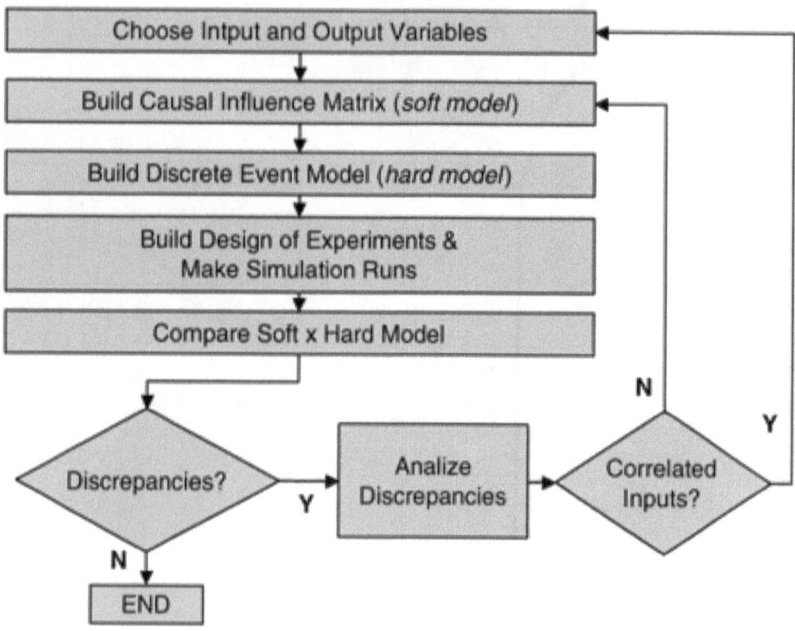

Figure 6-5. The Proposed Validation Technique

Note that, by this approach, we are supposing that if no more discrepancies exist, the model deserves higher confidence. Nevertheless, both the correlation matrix and the model could be wrong. Since this sort of thing has a very low probability to occur, the V&V technique should be complete when no discrepancies are found. Another issue arises with the choice of the number of variables the analyst considers. If the analysts tried to cover 100% of the variables, this procedure becomes unfeasible in practice, losing its principle of simplicity. According to G. A. Miller [50], good number of input and output variable is seven, plus or minus two.

Analysts should consider "critical variables" that correspond to critical system factors.

(Step 9) Prepare the Report

The analyst must be sure to include conclusions and recommendations about the subject models, the validation process, and future assessments of the models. The outline for this report is shown in Table 6-1.

Table 6-1. Reporting Outline

PARA NO.	TITLE	COMMENTS
1.	Overview	This includes general introductory information including, but not limited to, the following: • Model version • Description of analysis objective & approach • Specific resources used and personnel involved in performing the analysis • General comments on Soft Model Comparison
2.	Approach	What were you trying to achieve? State the agreed-upon objectives for the Soft Model Comparison. W hat did you do? Did you change the method? Include a description of the subject models, including identification of the particular version under assessment
2.1	Resources/References	List and number for referencing in the text such as the APN, environment data, and threat scenario.
3.	Procedure Results	This section records the results of all activity within the Execution Phase.

PARA NO.	TITLE	COMMENTS
3.1	Problem Definition	Briefly describe the problem definition agreed upon. Include the conference minutes documenting the problem definition. Additional comments and recommendations. Include specific recommended procedures, if applicable, and justification.
3.2	Independent Simulation	Summarize the results of the simulation. Outputs from the models during the simulation runs step (7), the comparison step (8), and the list of identified differences Include the preliminary report of the differences discovered as an attachment. Additional comments and recommendations. Include specific recommended procedures, if applicable, and justification.
3.3	Interactive Simulation Results	Describe the resolutions needed to bring the models into agreement. Summarize the output from the models following the interactive simulation step Include the causes and resolutions for all of the differences discovered in the models. Explain any redefinition of the problem that had to be done. Document the adjusting of parameters required that the simulated effects agree, even if the input parameters are different.

PARA NO.	TITLE	COMMENTS
4.0	Summary and Recommendations	This section provides highlights of the individual activities. It is a general summary with specific attention given to those areas that are particularly proficient or problematic. It also provides recommendations addressing the model and those areas that are particularly problematic.

Notes

[1] Fuzzy sets are sets whose elements have degrees of membership. Fuzzy sets were introduced simultaneously by Lotfi A. Zadeh and Dieter Klaua in 1965 as an extension of the classical notion of set. In classical set theory, the membership of elements in a set is assessed in binary terms according to a bivalent condition — an element either belongs or does not belong to the set. By contrast, fuzzy set theory permits the gradual assessment of the membership of elements in a set; this is described with the aid of a membership function valued in the real unit interval [0, 1].

[2] Artificial neural networks are composed of interconnecting artificial neurons (programming constructs that mimic the properties of biological neurons). Artificial neural networks may either be used to gain an understanding of biological neural networks, or for solving artificial intelligence problems without necessarily creating a model of a real biological system. The real, biological nervous system is highly complex: artificial neural network algorithms attempt to abstract this complexity and focus on what may hypothetically matter most from an information processing point of view.

[3] A Bayesian network, belief network or directed acyclic graphical model is a probabilistic graphical model that represents a set of random variables and their conditional dependencies via a directed acyclic graph (DAG). For example, a Bayesian network could represent the probabilistic relationships between diseases and symptoms. Given symptoms, the

network can be used to compute the probabilities of the presence of various diseases.

[4] Chaos theory is a field of study in mathematics, with applications in several disciplines including physics, economics, biology, and philosophy. Chaos theory studies the behavior of dynamical systems that are highly sensitive to initial conditions, an effect which is popularly referred to as the butterfly effect. Small differences in initial conditions (such as those due to rounding errors in numerical computation) yield widely diverging outcomes for chaotic systems, rendering long-term prediction impossible in general.

[5] In theoretical computer science, computational learning theory is a mathematical field related to the analysis of machine learning algorithms. Theoretical results in machine learning mainly deal with a type of inductive learning called supervised learning. In supervised learning, an algorithm is given samples that are labeled in some useful way. For example, the samples might be descriptions of mushrooms, and the labels could be whether or not the mushrooms are edible. The algorithm takes these previously labeled samples and uses them to induce a classifier. This classifier is a function that assigns labels to samples including the samples that have never been previously seen by the algorithm. The goal of the supervised learning algorithm is to optimize some measure of performance such as minimizing the number of mistakes made on new samples.

[6] A causal loop diagram (CLD) is a causal diagram that aids in visualizing how interrelated variables affect one another. The diagram consists of a set of nodes representing the variables connected together. The relationships between these variables, represented by arrows, can be labeled as positive or negative.

[7] In general usage, design of experiments (DOE) or experimental design is the design of any information-gathering exercises where variation is present, whether under the full control of the experimenter or not. However, in statistics, these terms are usually used for controlled experiments. In the design of experiments, the experimenter is often interested in the effect of some process or intervention (the "treatment") on some objects (the "experimental units"), which may be people, parts of people, groups of people, plants, animals, materials, etc. Design of experiments is thus a discipline that has very broad application across all the natural and social sciences.

[8] Analysis of Alternatives (AoA) is a cornerstone of Military Acquisition, and deliberately embodies the fair and competitive character of the United States business atmosphere. It is an effort to move from employing a single source to the exploration of multiple alternatives so agencies have a basis for funding the best possible projects in a rational, defensible manner considering risk and uncertainty. The AoA establishes and benchmarks metrics for Cost, Schedule, Performance (CSP) and Risk (CSPR) depending on military needs. The AoA also assesses critical technology elements (CTEs) associated with each proposed materiel solution, identified in the Initial Capabilities Document (ICD), including; technology maturity, integration risk, manufacturing feasibility, and, where necessary, technology maturation and demonstration needs. An AoA begins by establishing (or modifying) Key Performance Parameters (KPPs) metrics for each alternative. KPPs help compare the operational effectiveness, suitability, and life cycle costs of alternatives to satisfy the military need

[9] Assuming that it can be constructed in a meaningful way, the causal influence matrix provides the basis for several other measures that are of interest to the analyst: utility of instruments, disagreement and conflict between actors, and problem solving potential.

[10] The correlation matrix of n random variables $X_1, \ldots, X_n$ is the $n \times n$ matrix whose i, j entry is $\operatorname{corr}(X_i, X_j)$. If the measures of correlation used are product-moment coefficients, the correlation matrix is the same as the covariance matrix of the standardized random variables $X_i\ /\ \sigma\ (X_i)$ for $i = 1, \ldots, n$. This applies to both the matrix of population correlations (in which case "σ" is the population standard deviation), and to the matrix of sample correlations (in which case "σ" denotes the sample standard deviation).

[11] In statistics, a full factorial experiment is an experiment whose design consists of two or more factors, each with discrete possible values or "levels", and whose experimental units take on all possible combinations of these levels across all such factors. A full factorial design may also be called a fully crossed design. Such an experiment allows studying the effect of each factor on the response variable, as well as the effects of interactions between factors on the response variable.

Chapter 7. Structured analysis

Structured Analysis (SA) in software engineering and its allied technique, Structured Design (SD), are methods for analyzing and converting business requirements into specifications and ultimately, computer programs, hardware configurations and related manual procedures. Figure 7-1 shows an example of a structured analysis approach.

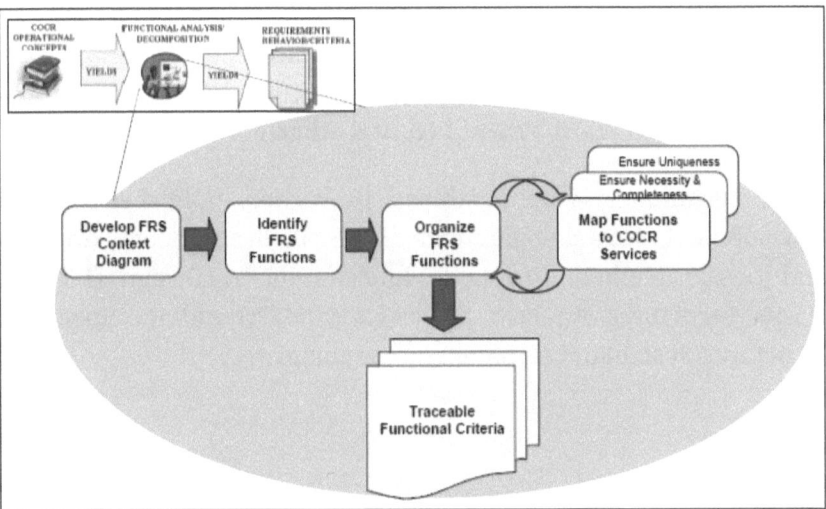

Figure 7-1. Example of a Structured Analysis approach [51].

Structured analysis and design techniques are fundamental tools of systems analysis, and developed from classical systems analysis of the 1960s and 1970s [52].

Objectives of Structured Analysis

Structured Analysis became popular in the 1980s and is still used by many. The analysis consists of interpreting the system concept (or real world) into data and control terminology, which one puts into data flow diagrams. The flow of data and control from bubble to data store to bubble can be very hard to track and the number of bubbles can get to be extremely large. One approach is to first define events from the outside world that require the system to react, then assign a bubble to that event, bubbles that need to interact are then connected until the system is defined. This can be rather overwhelming and so the bubbles are usually grouped into higher-level bubbles. Data Dictionaries[1] are needed to describe the data and command flows and a process specification is needed to capture the transaction/transformation information [53].

SA and SD were accompanied by notational methods including structure charts[2], data flow diagrams[3] and data model diagrams[4], of which there were many variations, including those developed by Tom DeMarco, Ken Orr, Larry Constantine, Vaughn Frick, Ed Yourdon, Steven Ward, Peter Chen, and others.

These techniques were combined in various published System Development Methodologies, including Structured Systems Analysis and Design Method, Profitable Information by Design (PRIDE), Nastec Structured Analysis & Design, SDM/70 and the Spectrum Structured system development methodology.

History

Structured analysis is part of a series of structured methods, which *"represent a collection of analysis, design, and programming techniques that were developed in response to the problems facing the software world from the 1960s to the 1980s. In this timeframe, most commercial programming was done in Cobol and Fortran, then C and BASIC. There was little guidance on "good" design and programming techniques, and there were no standard techniques for documenting requirements and designs. Systems where getting larger and more complex, and the information system development became harder and harder to do so"* [54]. As a way to help manage large and complex software.

Since the end 1960, multiple Structured Methods emerged [54]:

- Structured programming in circa 1967 with Edsger Dijkstra - "Go To Statement Considered Harmful"
- Niklaus Wirth Stepwise design in 1971
- Nassi–Shneiderman diagram in 1972
- Warnier/Orr diagram in 1974 - "Logical Construction of Programs"
- HIPO in 1974 - IBM Hierarchy input-process-output (though this should really be output-input-process)
- Structured Design around 1975 with Larry Constantine, Ed Yourdon and Wayne Stevens.
- Jackson Structured Programming in circa 1975 developed by Michael A. Jackson
- Structured Analysis in circa 1978 with Tom DeMarco, Yourdon, Gane & Sarson, McMenamin and Palmer.
- Structured Analysis and Design Technique (SADT) developed by Douglas T. Ross
- Yourdon Structured Method developed by Edward Yourdon.
- Structured Analysis and System Specification published in 1979 by Tom DeMarco.
- Structured Systems Analysis and Design Method (SSADM) first presented in 1983 developed by the UK Office of Government Commerce.
- IDEF0 based on SADT, developed by Douglas T. Ross in 1985 [55].
- Information Engineering in circa 1990 with Finkelstein and popularized by James Martin.

According to Hay (1999), *"information engineering was a logical extension of the structured techniques that were developed during the 1970's. Structured programming led to structured design, which in turn led to structured systems analysis. These techniques were characterized by their use of diagrams: structure charts for structured design, and data flow diagrams for structured analysis, both to aid in communication between users and developers, and to improve the analyst's and the designer's discipline. During the 1980's, tools began to appear which both automated the drawing of the* diagrams, and kept track of the things drawn in a data dictionary" [56]. After the example of computer-aided design and computer-aided manufacturing (CAD/CAM), the use of these tools was named Computer-aided software engineering (CASE).

Structured analysis topics
Single abstraction mechanism

Structured analysis typically creates a hierarchy employing a single abstraction mechanism. The structured analysis method can employ IDEF or Integration Definition (see Figure 7-2), is process driven, and starts with a purpose and a viewpoint. This method identifies the overall function and iteratively divides functions into smaller functions, preserving inputs, outputs, controls, and mechanisms necessary to optimize processes. Also known as a functional decomposition[5] approach, it focuses on cohesion within functions and coupling between functions leading to structured data [57].

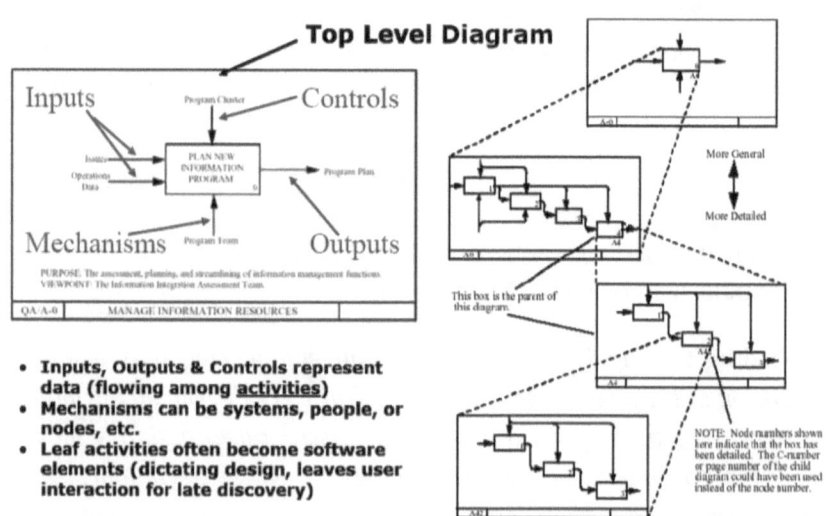

Figure 7-2. Image Structured Analysis example [57].

The functional decomposition of the structured method describes the process without delineating system behavior and dictates system structure in the form of required functions. The method identifies inputs and outputs as related to the activities. One reason for the popularity of structured analysis is its intuitive ability to communicate high-level processes and concepts, whether single system or enterprise levels. Discovering how objects might support functions for commercially prevalent object-oriented[6] development is unclear. In contrast to IDEF[7], the UML[8] is interface driven with multiple abstraction mechanisms useful in describing service-oriented architectures[9] (SOAs) [57].

Approach

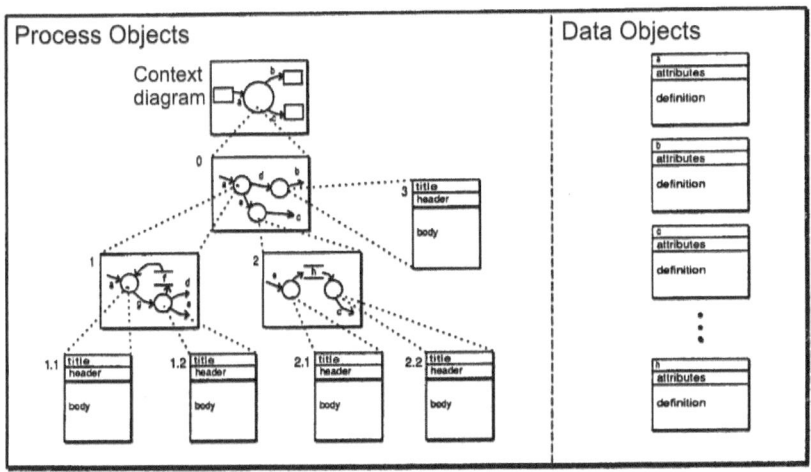

Figure 7-3. The structured analyze approach develops perspectives on both process objects and data objects [58].

Structured Analysis views a system from the perspective of the data flowing through it. Processes that transform the data flows describe the function of the system. Structured analysis takes advantage of information hiding through successive decomposition (or top down) analysis. This allows attention to be focused on pertinent details and avoids confusion from looking at irrelevant details. As the level of detail increases, the breadth of information is reduced. The result of structured analysis is a set of related graphical diagrams, process descriptions, and data definitions. They describe the transformations that need to take place and the data required to meet a system's functional requirements[10] [58]. Figure 7-3 depicts this approach.

De Marco's approach [59] consists of the following objects (see figure) [58]:

- Context diagram
- Data flow diagram,
- process specifications, and
- a data dictionary,

Hereby the Data flow diagrams (DFDs) are directed graphs. The arcs represent data, and the nodes (circles or bubbles) represent processes that transform the data. A process can be further decomposed to a more detailed DFD, which shows the subprocesses, and data flows within it. The subprocesses can in turn be decomposed further with another set of DFDs until their functions can be easily understood. Functional primitives are processes that do not need to be decomposed further. Functional primitives are described by a process specification (or mini-spec). The process specification can consist of pseudo-code, flowcharts, or structured English. The DFDs model the structure of the system as a network of interconnected processes composed of functional primitives. The data dictionary is a set of entries (definitions) of data flows, data elements, files, and databases. The data dictionary entities are partitioned in a top down manner. They can be referenced in other data dictionary entries and in data flow diagrams [58].

Context diagram

Context diagrams are diagrams that represent the actors outside a system that could interact with that system [60]. This diagram is the highest-level view of a system, similar to Block Diagram, showing a, possibly software-based, system as a whole and its inputs and outputs from/to external factors.

This type of diagram according to Kossiakoff usually "*pictures the system at the center, with no details of its interior structure, surrounded by all its interacting systems, environment and activities. The objective of a system context diagram is to focus attention on external factors and events that should be considered in developing a complete set of system requirements and constraints*" [60]. System context diagram are related to Data Flow Diagram, and show the interactions between a system and other actors with which the system is designed to face. System context diagrams can be helpful in understanding the context in which the system will be part of software engineering (see Figure 7-4).

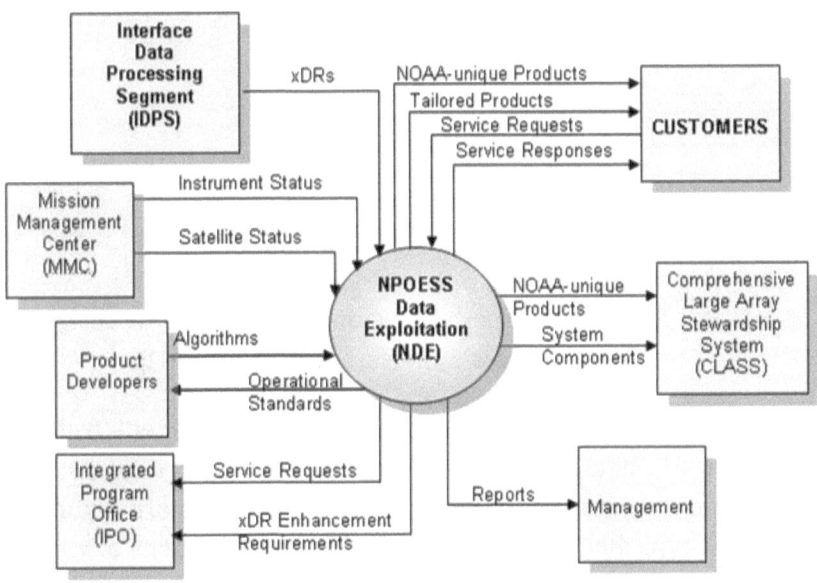

Figure 7-4. Example of a System context diagram [61].

Data dictionary

A data dictionary or *database dictionary* is a file that defines the basic organization of a database [62]. A database dictionary contains a list of all files in the database, the number of records in each file, and the names and types of each data field. Most database management systems keep the data dictionary hidden from users to prevent them from accidentally destroying its contents. Data dictionaries do not contain any actual data from the database, only bookkeeping information for managing it. Without a data dictionary, however, a database management system cannot access data from the database [62].

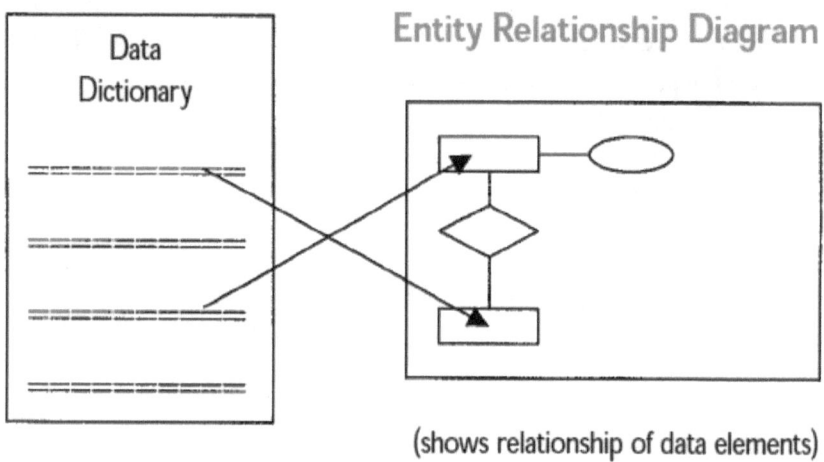

Figure 7-5. Entity relationship diagram, essential for the design of database tables, extracts, and metadata [62].

Database users and application developers can benefit from an authoritative data dictionary document that catalogs the organization, contents, and conventions of one or more databases [63]. This typically includes the names and descriptions of various tables and fields in each database, plus additional details, like the type and length of each data element. There is no universal standard as to the level of detail in such a document, but it is primarily a distillation of metadata[11] about database structure, not the data itself. A data dictionary document also may include further information describing how data elements are encoded. One of the advantages of well-designed data dictionary documentation is that it helps to establish consistency throughout a complex database, or across a large collection of federated databases[12] [64].

Data Flow Diagrams

A Data Flow Diagram (DFD) is a graphical representation of the "flow" of data through an information system. It differs from the system flowchart as it shows the flow of data through processes instead of hardware. Data flow diagrams were invented by Larry Constantine, developer of structured design, based on Martin and Estrin's "data flow graph" model of computation [65].

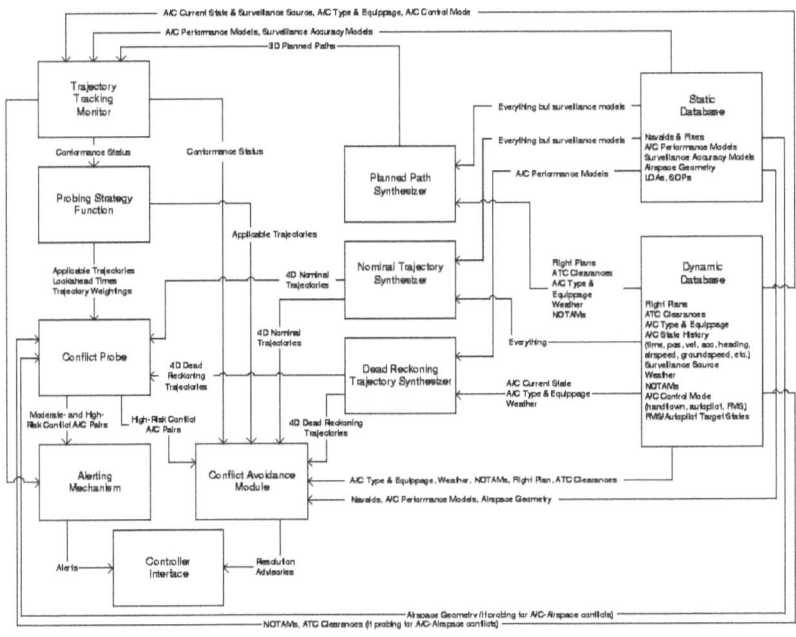

Figure 7-6. Data Flow Diagram example [66].

It is common practice to draw a System Context Diagram first, which shows the interaction between the system and outside entities. The DFD is designed to show how a system is divided into smaller portions and to highlight the flow of data between those parts. This context-level Data flow diagram is then "exploded" to show more detail of the system being modeled.

Data flow diagrams (DFDs) are one of the three essential perspectives of Structured Systems Analysis and Design Method

(SSADM). The sponsor of a project and the end users will need to be briefed and consulted throughout all stages of a system's evolution. With a data flow diagram, users are able to visualize how the system will operate, what the system will accomplish, and how the system will be implemented. The old system's data flow diagrams can be drawn up and compared with the new system's data flow diagrams to draw comparisons to implement a more efficient system. Data flow diagrams can be used to provide the end user with a physical idea of where the data they input ultimately has an effect upon the structure of the whole system from order to dispatch to recook. How any system is developed can be determined through a dataflow diagram.

Structure Chart

A Structure Chart (SC) is a chart (see Figure 7-7), that shows the breakdown of the configuration system to the lowest manageable levels [67]. This chart is used in structured programming to arrange the program modules in a tree structure. Each module is represented by a box that contains the name of the modules. The tree structure visualizes the relationships between the modules [68].

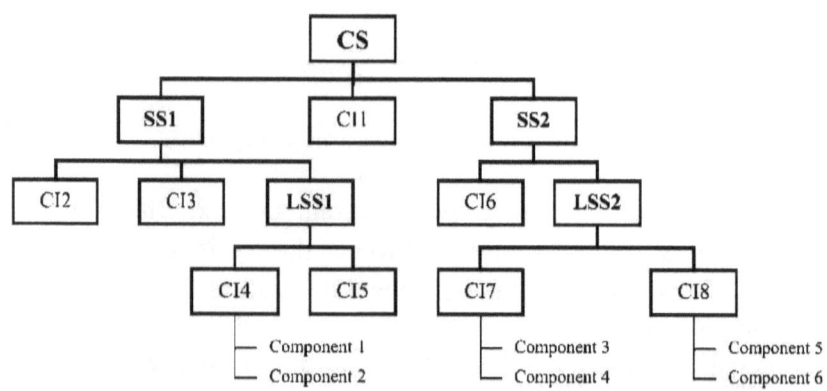

Figure 7-7. A Configuration System Structure Chart [67].

In structured analysis, structure charts are used to specify the high-level design, or architecture, of a computer program. As a design tool, they aid the programmer in dividing and conquering a large software problem, that is, recursively breaking a problem down into parts that are small enough to be understood by a human brain. The process is called top-down design, or functional decomposition. Programmers use a structure chart to build a program in a manner similar to how an architect uses a blueprint to build a house. In the design stage, the chart is drawn and used as a way for the client and the various software designers to communicate. During the actual building of the program (implementation), the chart is continually referred to as the master plan [69].

Structured Design

Structured Design (SD) is concerned with the development of modules and the synthesis of these modules in a so called "module hierarchy" [70]. In order to design optimal module structure and interfaces two principles are crucial:

Cohesion which is "concerned with the grouping of functionally related processes into a particular module" [58], and

Coupling relates to "the flow of information, or parameters, passed between modules. Optimal coupling reduces the interfaces of modules and the resulting complexity of the software" [58].

Page-Jones (1980) has proposed his own approach, which consists of three main objects: structure charts, module specifications and a data dictionary [70]. The structure chart aims to show *"the module hierarchy or calling sequence relationship of modules. There is a module specification for each module shown on the structure chart. The module specifications can be composed of pseudo-code or a program design language. The data dictionary is like that of structured analysis. At this stage in the software development lifecycle, after analysis and design have been performed, it is possible* to automatically generate data type declarations" [71] and procedure or subroutine templates [58].

Structured query language

The structured query language (SQL) is a standardized language for querying information from a database. SQL was first introduced as a commercial database system in 1979 and has since been the favorite query language for database management systems running on minicomputers and mainframes. Increasingly, however, SQL is being supported by PC database systems because it supports distributed databases (see definition of distributed database). This enables several users on a computer network to access the same database simultaneously. Although there are different dialects of SQL, it is nevertheless the closest thing to a standard query language that currently exists [62].

Criticisms

Problems with data flow diagrams have been [53]:

- choosing bubbles appropriately,
- partitioning those bubbles in a meaningful and mutually agreed upon manner,
- the size of the documentation needed to understand the Data Flows,
- still strongly functional in nature and thus subject to frequent change,
- though "data" flow is emphasized, "data" modeling is not, so there is little understanding of just what the subject matter of the system is about, and
- not only is it hard for the customer to follow how the concept is mapped into these data flows and bubbles, it has also been very hard for the designers who must shift the DFD organization into an implementable format

Notes

[1] A data dictionary, or metadata repository, as defined in the IBM Dictionary of Computing, is a "centralized repository of information about data such as meaning, relationships to other data, origin, usage, and format"(ACM, IBM Dictionary of Computing, 10th edition, 1993). The term may have one of several closely related meanings pertaining to databases and database management systems (DBMS): a document describing a database or collection of databases; an integral component of a DBMS that is required to determine its structure; a piece of middleware that extends or supplants the native data dictionary of a DBMS

[2] A Structure Chart (SC) in software engineering and organizational theory is a chart, which shows the breakdown of a system to its lowest manageable levels.

[3] A data flow diagram (DFD) is a graphical representation of the "flow" of data through an information system, modeling its process aspects. Often they are a preliminary step used to create an overview of the system which can later be elaborated.

[4] A data model in software engineering is an abstract model that documents and organizes the business data for communication between team members and is used as a plan for developing applications, specifically how data is stored and accessed.

[5] Functional decomposition refers broadly to the process of resolving a functional relationship into its constituent parts in such a way that the original function can be reconstructed (i.e., recomposed) from those parts by function composition. In general, this process of decomposition is undertaken either for the purpose of gaining insight into the identity of the constituent components (which may reflect individual physical processes of interest, for example), or for the purpose of obtaining a compressed representation of the global function, a task which is feasible only when the constituent processes possess a certain level of modularity (i.e., independence or non-interaction).

[6] Object-oriented programming (OOP) is a programming paradigm using "objects" – data structures consisting of data fields and methods together with their interactions – to design applications and computer programs. Programming techniques may include features such as data abstraction, encapsulation, messaging, modularity, polymorphism, and inheritance. Many modern programming languages now support OOP, at least as an option.

⁷ IDEF, an abbreviation of Integration Definition, refers to a family of modeling languages in the field of systems and software engineering. They cover a wide range of uses, from functional modeling to data, simulation, object-oriented analysis/design and knowledge acquisition. These "definition languages" were developed under funding from U.S. Air Force and although still most commonly used by them, as well as other military and Department of Defense (DoD) agencies, are in the public domain. The most-widely recognized and used components of the IDEF family are IDEF0, a functional modeling language building on SADT, and IDEF1X, which addresses information models and database design issues.

⁸ Unified Modeling Language (UML) is a standardized general-purpose modeling language in the field of object-oriented software engineering. The standard is managed, and was created, by the Object Management Group. UML includes a set of graphic notation techniques to create visual models of object-oriented software-intensive systems.

⁹ Service-orientation is a design paradigm to build computer software in the form of services. Like other design paradigms (e.g. object-orientation), service-orientation provides a governing approach to automate business logic as distributed systems. What distinguishes service-orientation is its set of design principles to ensure the manner in which it carries out the separation of concerns in the software. A service-oriented architecture (SOA) is governed by these principles. Applying service-orientation results in units of software partitioned into operational capabilities, each designed to solve an individual concern. These units qualify as services.

¹⁰ In software engineering, a functional requirement defines a function of a software system or its component. A function is described as a set of inputs, the behavior, and outputs (see also software). Functional requirements may be calculations, technical details, data manipulation and processing and other specific functionality that define what a system is supposed to accomplish. Behavioral requirements describing all the cases where the system uses the functional requirements are captured in use cases. Functional requirements are supported by non-functional requirements (also known as quality requirements), which impose constraints on the design or implementation (such as performance requirements, security, or reliability). Generally, functional requirements are expressed in the form "system must do <requirement>", while non-functional requirements are "system shall be <requirement>". The plan for implementing functional requirements is detailed in the system design.

The plan for implementing non-functional requirements is detailed in the system architecture.

[11] The term metadata is an ambiguous term that is used for two fundamentally different concepts (types). Although the expression "data about data" is often used, it does not apply to both in the same way. Structural metadata, the design and specification of data structures, cannot be about data, because at design time the application contains no data. In this case the correct description would be "data about the containers of data". Descriptive metadata, on the other hand, is about individual instances of application data, the data content. In this case, a useful description (resulting in a disambiguating neologism) would be "data about data contents" or "content about content" thus metacontent. Descriptive, Guide and the National Information Standards Organization concept of administrative metadata are all subtypes of metacontent.

[12] A federated database system is a type of meta-database management system (DBMS), which transparently integrates multiple autonomous database systems into a single federated database. The constituent databases are interconnected via computer network, and may be geographically decentralized. Since the constituent database systems remain autonomous, a federated database system is a contrastable alternative to the (sometimes daunting) task of merging together several disparate databases. A federated database, or virtual database, is the fully integrated, logical composite of all constituent databases in a federated database system.

Chapter 8. Data flow diagram

A **data flow diagram** (**DFD**) is a graphical representation of the "flow" of data through an information system, modeling its *process* aspects, as depicted in Figure 8-1. Often they are a preliminary step used to create an overview of the system that can later be elaborated [72]. We can also use DFDs for the visualization of data processing (structured design).

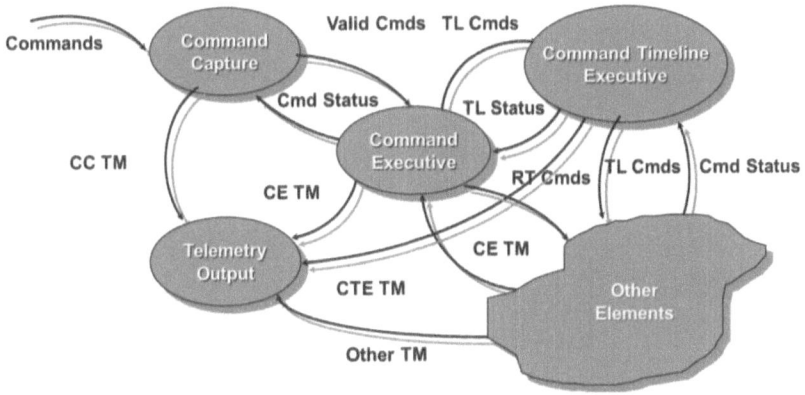

Figure 8-1. Data flow diagram example [66].

A DFD shows what kinds of data will be input to and output from the system, where the data will come from and go to, and where the data will be stored. It does not show information about the timing of processes, or information about whether processes will operate in sequence or in parallel (which is shown on a flowchart).

Overview

It is common practice to draw the context-level DFD first, that shows the interaction between the system and external agents that act as data sources and data sinks. On the context diagram, the system's interactions with the outside world are modeled purely in terms of data flows across the *system boundary*. The context diagram shows the entire system as a single process, and gives no clues as to its internal organization.

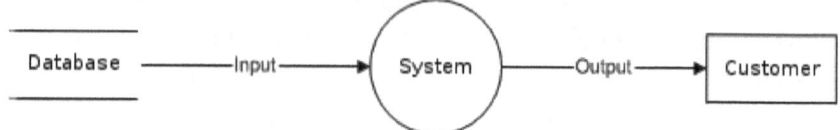

Figure 8-2(a). Data flow diagram example.

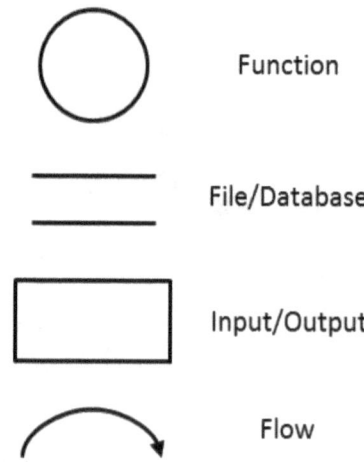

Figure 8-2(b). Data flow diagram - Yourdon/DeMarco notation.

This context-level DFD is next "exploded", to produce a Level 0 DFD that shows some of the detail of the system being modeled. The Level 0 DFD shows how the system is divided into sub-systems (processes), each of which deals with one or more of the data flows to or from an external agent, and which together provide all of the functionality of the system as a whole. It also identifies internal data stores that must be present in order for the system to do its job, and shows the flow of data between the various parts of the system.

Data flow diagrams were proposed by Larry Constantine, the original developer of structured design [65], based on Martin and Estrin's "data flow graph" model of computation [73].

Data flow diagrams (DFDs) are one of the three essential perspectives of the structured-systems analysis and design method SSADM. The sponsor of a project and the end users will need to be briefed and consulted throughout all stages of a system's evolution. With a data flow diagram, we are able to visualize how the system will operate, what the system will accomplish, and how the system will be implemented. We can draw up the old system's dataflow diagrams and compare them with the new system's data flow diagrams to draw comparisons to implement a more efficient system. Data flow diagrams can be used to provide the end user with a physical idea of where the data they input ultimately has an effect upon the structure of the whole system from order to dispatch to report. How any system is developed can be determined through a data flow diagram.

In the course of developing a set of *leveled* data flow diagrams, the analyst/designers is forced to address how the system may be decomposed into component subsystems, and to identify the transaction data[1] in the data model.

There are different notations to draw data flow diagrams (Yourdon & Coad and Gane & Sarson [74]), defining different visual representations for processes, data stores, data flow, and external entities [75].

Advantages of DFD

- Graphic technique is superb and simple.
- System boundaries are well described.
- Each part of data can be represented by different level of details [65].

Notations of DFDs
Data Flow

- Data flows from source to target except from external entity to data store and vice-versa.
- However, data flow can be possible from process to process, external entity to process and vice-versa, into and out of store from process.
- Indicated by arrowhead symbol with the name of the flow above arrow [75].

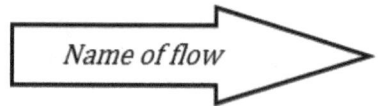

Figure 8-3. Data flow Notation.

Processes

- Process name indicates what action to be taken on data.
- Indicated by circle with a process name in it.
- Input data flow is transformed to output data flow [75].

Figure 8-4. Process Notation.

Data Stores

- It stores the information of data flow like files, database, etc.
- Indicated by parallel lines with a data store name above [75].

Figure 8-5. Data Stores Notation.

External Entities

- Represents external data processing units other than regular data flow.
- Depends on out of the system boundaries [75].

Resource store

- Represents physical material flow from source to target.
- They are usually restricted to high-level diagrams [75].

Steps
Types of DFD

Following four types are used in development project:

Physical DFDs
- Project scope and current system is well defined.
- Simple prototype2 model is drawn for estimation of purposes and to define a basic scope.
- Later a more complex design can be drawn for business purpose.
- Uses Level 1 and leveled steps.

Logical DFDs
- Drawn from current Physical DFD.
- Indicates the basic underlying functionality.
- No constraints are imposed.
- Uses Logical steps.

Business System Options
- Use to form the base of the required business system.
- Uses Level 1 steps.

Required DFDs
- Indicates the required system and then developed by using DFDs to satisfy the selected business system option.
- Uses Leveled steps.

Rules of DFD

External Entities
- Information within a system is obtained from or given to external entity.
- Crossing of data flow lines can be prevented by multiple similar external entities. When it happens, a strip is drawn across left hand corner.

Processes
- Processes should not be named without understanding their role.
- User of the DFD must understand the meaning of Description.

Data Flows
- All except bottom level diagrams can use Double headed arrows to show 2-way flows.
- In addition, a hierarchy of Data Flows can be constructed where Data Flow at each level divides into lower levels.

Data Stores
- Data Store should be given any number preceded by a reference letter as follows:
  1. 'D': Permanent computer file.
  2. 'M': Manual file.
  3. 'T': Transient Store.

- Data Stores can appear several times in DFD in case of complex diagram and should be indicated by double vertical bar on their left hand edge.

Context Diagrams

We use Context Diagrams to clear the scope of the entire system under investigation. A single process resembles context diagram, connected to external entities by resource flows and data flows. Hence, this diagram can be used to show the interfaces between system and external entities. Thus, context diagram aims at system boundary and clears the scope of analysis precisely.

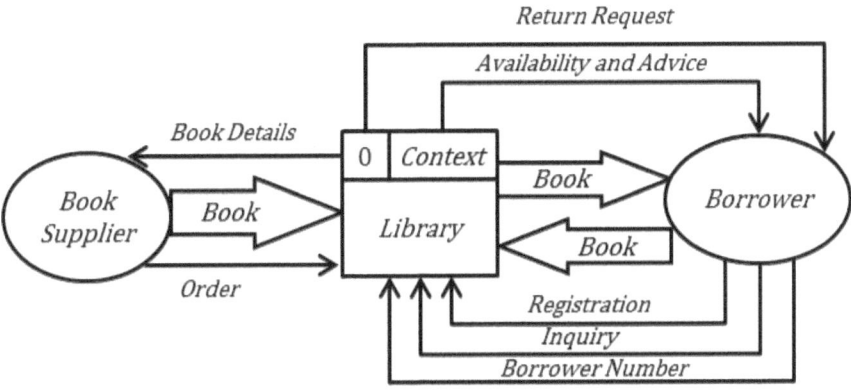

Figure 8-6. Context diagram example.

Level 1 Diagrams

Level 1 Diagram exploits the behavioral nature of system under investigation. There should be only one representation of Level 1 Diagram as with Context diagram. Since it depicts whole of the system, obtaining initial position can be difficult. Level 1 processes do not have any kind of formulation and should describe only behavioral elements of systems. Usually, we use higher-level processes to represent Level 1 Diagram. The outline of a process box, representing boundaries of system, surrounds Level 1 Diagram.

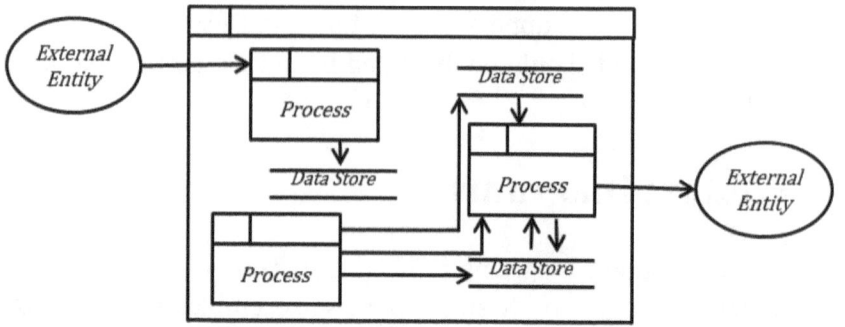

Figure 8-7. Level 1 diagram example.

The analyst should draw data flow diagrams in several nested layers. We can expand a single process node on a high-level diagram to show a more detailed data flow diagram. Draw the context diagram first, followed by various layers of data flow diagrams.

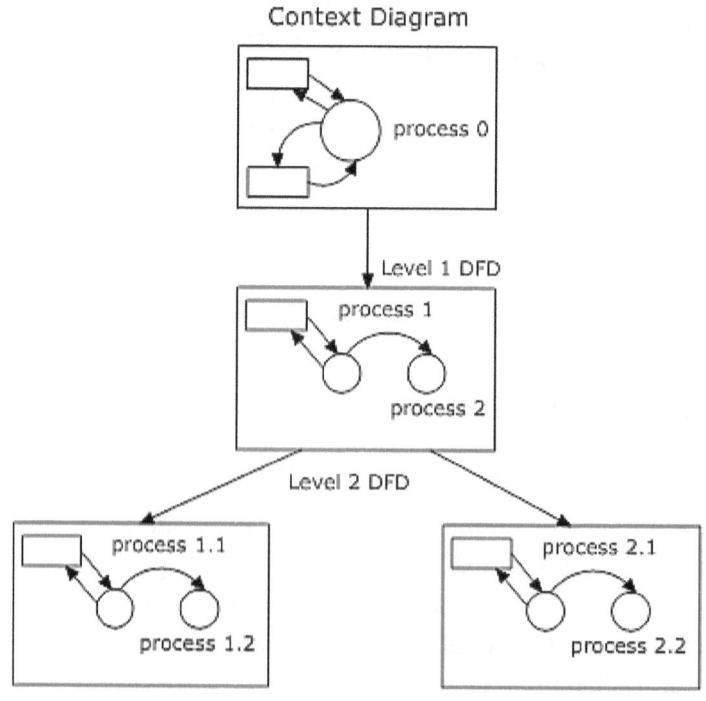

Figure 8-8. Nesting of data flow layers

Resource Flow Analysis

Based on flow of physical objects, this approach of Resource Flow Analysis is good for the system having large flow of goods. We can track physical resources in the following manner:

- Arriving within the boundaries of the system.
- Action occurrence.
- Exit from the system.

Important thing that we should consider behind this analysis is that information follows the same path as the physical objects.

Document Flow Analysis

By this approach, we can appropriately show the principle where information flowing in the form of documents or computer input and output used by the system under investigation. We can perform this analysis in following steps:

- List down the major documents and their sources and recipients.
- Identification of other major information.
- Draw the Document Flow Diagram.
- System Boundary should be added.

Organizational Structure Analysis

- Initiate the analysis by identification of main roles within an organization.
- Then the relevant functional areas are identified from the key processes.
- Then Discrete process can be identified by more detailing and analyzing.
- Based on this result, flow of information between them and external entities can be tracked and added to diagram.

Notes

[1] Transaction data is data describing an event (the change as a result of a transaction) and is usually described with verbs. Transaction data always has a time dimension, a numerical value and refers to one or more objects (i.e. the reference data).

[2] A prototype is an early sample or model built to test a concept or process or to act as a thing to be replicated or learned from. The word prototype derives from the Greek πρωτότυπον (prototypon), "primitive form", neutral of πρωτότυπος (prototypos), "original, primitive", from πρῶτος (protos), "first" and τύπος (typos), "impression".

Chapter 9. Jackson Structured Programming

Jackson Structured Programming or **JSP** is a method for structured programming based on correspondences between data stream structure and program structure. JSP structures programs and data in terms of sequences, iterations and selections, and as a consequence it is applied when designing a program's detailed control structure, below the level where object-oriented methods become important [76] [77].

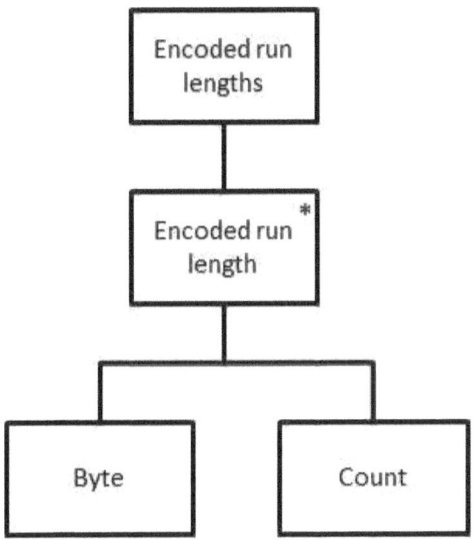

Figure 9-1. Example of a JSP diagram.

Introduction

Michael A. Jackson (not the late singer, Michael J. Jackson) originally developed JSP in the 1970s. He documented the system in his 1975 book *Principles of Program Design* [78]. Jackson's aim was to make COBOL[1] batch file processing programs easier to modify and maintain, but we can use the method to design programs for any programming language that has structured control constructs, languages such as C[2], Java[3] and Perl[4]. Despite its age, JSP is still in use and is supported by diagramming tools such as Microsoft's Visio and CASE tools such as Jackson Workbench[5] [79].

Jackson Structured Programming was seen by many as related [80] to Warnier Structured Programming[6] [81], but the latter method focused almost exclusively on the structure of the output stream. JSP and Warnier's method both structure programs and data using only sequences, iterations and selections, so they essentially create programs that are parsers[7] for regular expressions[8] which simultaneously match the program's input and output data streams.

Because JSP focuses on the existing input and output data streams, designing a program using JSP is claimed to be more straightforward than with other structured programming methods, avoiding the leaps of intuition needed to successfully program using methods such as top-down decomposition [82].

Another consequence of JSP's focus on data streams is that it creates program designs with a very different structure to the kind created by the stepwise refinement methods of Wirth and Dijkstra[9]. One typical feature of the structure of JSP programs is that they have several input operations distributed throughout the code in contrast to programs designed using stepwise refinement, which tend to have only one input operation. Jackson illustrates this difference in Chapter 3 of *Principles of Program Design* [78]. He presents two versions of a program; one designed using JSP, the other using "traditional" methods.

Structural equivalent

The JSP version of the program is structurally equivalent to

```
String line;

line = in.readLine();
while (line != null) {
    int count = 0;
```

```
    String firstLineOfGroup = line;

    while (line != null &&
line.equals(firstLineOfGroup)) {
        count++;
        line = in.readLine();
    }
    System.out.println(firstLineOfGroup + " " +
count);
}
```

and the traditional version of the program is equivalent to

```
String line;

int count = 0;
String firstLineOfGroup = null;
while ((line = in.readLine()) != null) {
    if (firstLineOfGroup == null
            || !line.equals(firstLineOfGroup)) {
        if (firstLineOfGroup != null) {
            System.out.println(firstLineOfGroup + " "
+ count);
        }
        count = 0;
        firstLineOfGroup = line;
    }
    count++;
}
if (firstLineOfGroup != null) {
    System.out.println(firstLineOfGroup + " " +
count);
}
```

Jackson criticizes the traditional version, claiming that it hides the relationships that exist between the input lines, compromising the program's understandability and maintainability by, for example, forcing the use of a special case for the first line and forcing another special case for a final output operation.

The method

JSP uses semi-formal steps to capture the existing structure of a program's inputs and outputs in the structure of the program itself.

The intent is to create programs that are easy to modify over their lifetime. Jackson's major insight was that requirement changes are usually minor tweaks to the existing structures. For a program constructed using JSP, the inputs, the outputs, and the internal structures of the program all match, so small changes to the inputs and outputs should translate into small changes to the program.

1. JSP structures programs in terms of four component types:
2. fundamental operations
3. sequences
4. iterations
5. selections

The method begins by describing a program's inputs in terms of the four fundamental component types. It then goes on to describe the program's outputs in the same way. Each input and output is modeled as a separate Data Structure Diagram[10] (DSD). To make JSP work for compute-intensive applications, such as digital signal processing (DSP) it is also necessary to draw algorithm structure diagrams, which focus on internal data structures rather than input and output ones.

The input and output structures are then unified or merged into a final program structure, known as a Program Structure Diagram[11] (PSD). This step may involve the addition of a small amount of high-level control structure to marry up the inputs and outputs. Some programs process all the input before doing any output, whilst others read in one record, write one record and iterate. We have to capture such approaches in the PSD.

Next, we implement the PSD, which is language neutral, in a programming language. JSP is geared towards programming at the level of control structures, so the implemented designs use just primitive operations, sequences, iterations and selections. JSP is not used to structure programs at the level of classes and objects, although it can helpfully structure control flow within a class's methods.

JSP uses a diagramming notation to describe the structure of inputs, outputs and programs, with diagram elements for each of the fundamental component types.

The programmer draws a simple operation as a box.

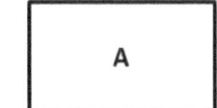

Figure 9-2. An operation

A sequence of operations is represented by boxes connected with lines. In the example below, operation A consists of the sequence of operations B, C and D.

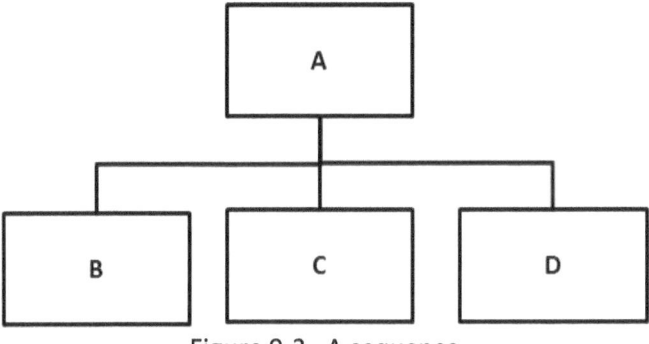

Figure 9-3. A sequence

An iteration is again represented with joined boxes. In addition, the iterated operation has a star in the top right corner of its box. In the example below, operation A consists of an iteration of zero or more invocations of operation B.

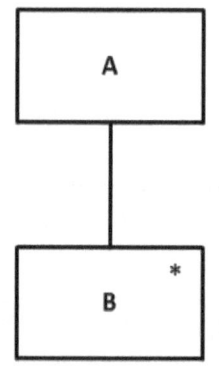

Figure 9-4. An iteration

Selection is similar to a sequence, but with a circle drawn in the top right hand corner of each optional operation. In the example, operation A consists of one and only one of operations B, C or D.

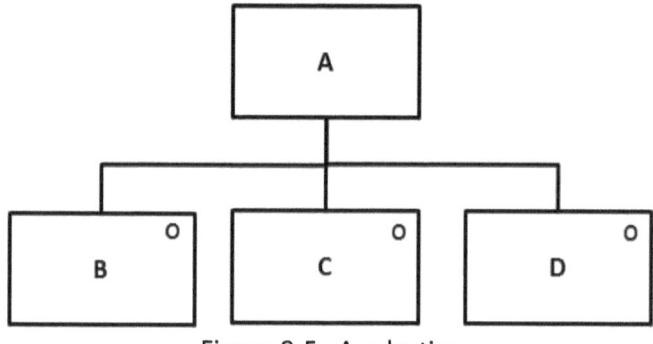

Figure 9-5. A selection

A worked example

As an example, here is how a programmer would design and code a run length encoder using JSP.

A run length encoder is a program that takes as its input a stream of bytes. It outputs a stream of pairs consisting of a byte along with a count of the byte's consecutive occurrences. Run-length encoders are often used for crudely compressing bitmaps.

With JSP, the first step is to describe the structure of a program's inputs. A run-length encoder has only one input, a stream of bytes that can be viewed as zero or more *runs*. Each run consists of one or more bytes of the same value. This is represented by the following JSP diagram.

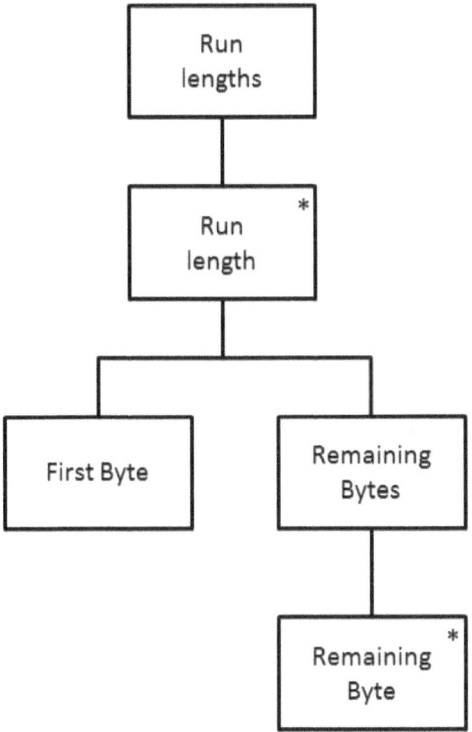

Figure 9-6. The run-length encoder input

The second step is to describe the structure of the output. The run-length encoder output can be described as zero or more pairs, each pair consisting of a byte and its count. In this example, the count will also be a byte.

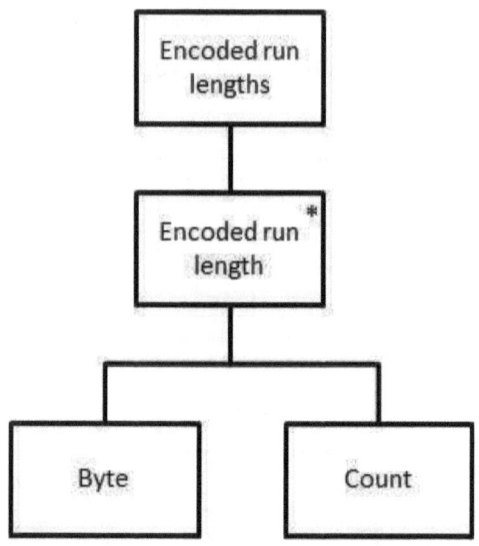

Figure 9-7. The run-length encoder output

The next step is to describe the correspondences between the operations in the input and output structures.

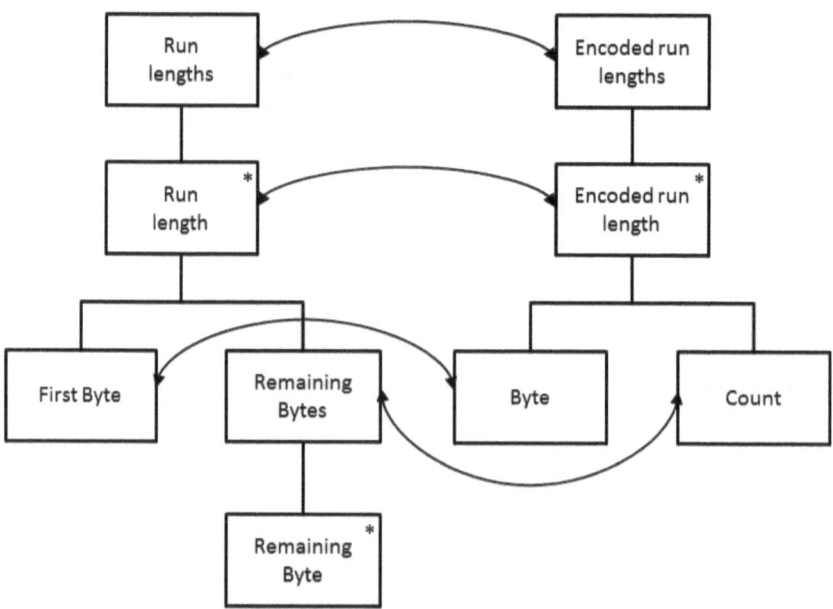

Figure 9-8. The correspondences between the run-length encoder's inputs and its outputs

It is at this stage that the astute programmer may encounter a *structure clash*, in which there is no obvious correspondence between the input and output structures. If we find a structure clash, we usually resolve it by splitting the program into two parts, using an intermediate data structure to provide a common structural framework with which the two program parts can communicate. The two programs parts are often implemented as processes or co-routines.

In this example, there is no structure clash, so the two structures can be merged to give the final program structure.

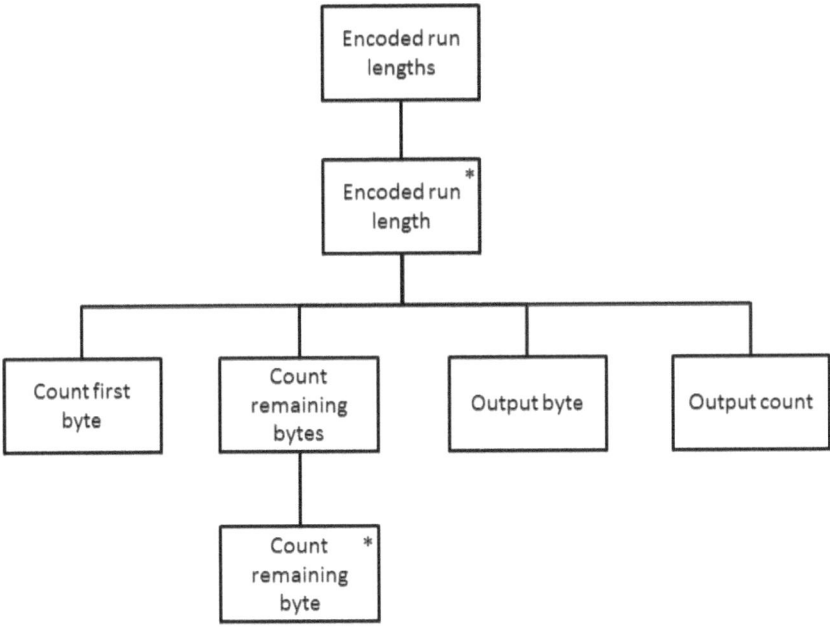

Figure 9-9. The run-length encoder program structure

At this stage, we can flesh out the program by hanging various primitive operations off the elements of the structure. Primitives who suggest themselves are

1. read a byte
2. remember byte

3. set counter to zero
4. increment counter
5. output remembered byte
6. output counter

The iterations also have to be fleshed out. They need conditions added. Suitable conditions would be

1. while there are more bytes
2. while there are more bytes and this byte is the same as the run's first byte and the count will still fit in a byte

If we put all this together, we can convert the diagram and the primitive operations into C, maintaining a one-to-one correspondence between the code and the operations and structure of the program design diagram.

```c
#include <stdio.h>
#include <stdlib.h>

int main(int argc, char *argv[])
{
    char c;

    c = getchar();
    while (c != EOF) {
        char count = 1;

        char first_byte = c;

        c = getchar();

        while (c != EOF && c == first_byte && count < 255)
{
            count++;
            c = getchar();
        }

        putchar(first_byte);
        putchar(count);
    }
    return EXIT_SUCCESS;
}
```

This method will work only when translation from input to output is equivalent to a context-free grammar.

Notes

[1] COBOL is one of the oldest programming languages. Its name is an acronym for COmmon Business-Oriented Language, defining its primary domain in business, finance, and administrative systems for companies and governments.

[2] C is a general-purpose computer programming language developed between 1969 and 1973 by Dennis Ritchie, at the Bell Telephone Laboratories, for use with the UNIX operating system. Although C was designed for implementing system software, it is also widely used for developing portable application software. C is one of the most widely used programming languages of all time and there are very few computer architectures for which a C compiler does not exist. C has greatly influenced many other popular programming languages, most notably C++, which began as an extension to C.

[3] Java is a programming language originally developed by James Gosling at Sun Microsystems (which is now a subsidiary of Oracle Corporation) and released in 1995 as a core component of Sun Microsystems' Java platform. The language derives much of its syntax from C and C++ but has a simpler object model and fewer low-level facilities. Java applications are typically compiled to bytecode (class file) that can run on any Java Virtual Machine (JVM) regardless of computer architecture. Java is a general-purpose, concurrent, class-based, object-oriented language that is specifically designed to have as few implementation dependencies as possible.

[4] Perl is a high-level, general-purpose, interpreted, dynamic programming language. Perl was originally developed by Larry Wall in 1987 as a general-purpose UNIX scripting language to make report processing easier. Perl borrows features from other programming languages including C, shell scripting (sh), AWK, and sed.

[5] The KCSL Jackson Workbench is a suite of modern CASE tools that support the Jackson Structured Programming (JSP) and Jackson System Development (JSD) software development methods devised by Michael A. Jackson. The tools that comprise the Jackson Workbench can operate independently or as an integrated Workbench. They provide direct

replacements for legacy tools such as PDF, JSP Tool, JSP-COBOL and JSP-MACRO.

⁶ A Warnier/Orr diagram (also known as a logical construction of a program/system) is a kind of hierarchical flowchart that allows the description of the organization of data and procedures. Jean-Dominique Warnier (France) and Kenneth Orr (United States) initially developed them. This method aids the design of program structures by identifying the output and processing results and then working backwards to determine the steps and combinations of input needed to produce them. The simple graphic method used in Warnier/Orr diagrams makes the levels in the system evident and the movement of the data between them vivid.

⁷ In computer science and linguistics, parsing, or, more formally, syntactic analysis, is the process of analyzing a text, made of a sequence of tokens (for example, words), to determine its grammatical structure with respect to a given (more or less) formal grammar. Parsing is also a linguistic term, especially in reference to how phrases are divided up in garden path sentences—a grammatically correct sentence that starts in such a way that the readers' most likely interpretation will be incorrect; they are lured into an improper parse that turns out to be a dead end.

⁸ In computing, a regular expression provides a concise and flexible means for "matching" (specifying and recognizing) strings of text, such as particular characters, words, or patterns of characters. Abbreviations for "regular expression" include "regex" and "regexp". The concept of regular expressions was first popularized by utilities provided by Unix distributions

⁹ Edsger Wybe Dijkstra (May 11, 1930 – August 6, 2002) was a Dutch computer scientist, who received the 1972 Turing Award for fundamental contributions to developing programming languages, and was the Schlumberger Centennial Chair of Computer Sciences at The University of Texas at Austin from 1984 until 2000. While he had programmed extensively in machine code in the 1950s, he was known for his low opinion of the GOTO statement in computer programming. He wrote a paper in 1965, culminating in the 1968 article "A Case against the GO TO Statement", which is regarded as a major step towards the widespread deprecation of the GOTO statement and its effective replacement by structured control constructs, such as the while loop. This article was retitled by editor Niklaus Wirth to "Go To Statement Considered Harmful",

which introduced the phrase "considered harmful" in computing. This methodology was also called structured programming, the title of his 1972 book, coauthored with C. A .R. Hoare and Ole-Johan Dahl. Dijkstra also strongly opposed the teaching of BASIC.

[10] A Data Structure Diagram (DSD) is a data model used to describe conceptual data models by providing graphical notations, which document entities and their relationships, and the constraints that binds them. The basic graphic elements of DSDs are boxes, representing entities, and arrows, representing relationships. Data structure diagrams are most useful for documenting complex data entities.

[11] A PSD called the Nassi–Shneiderman diagram (NSD) in computer programming is a graphical design representation for structured programming. This type of diagram was developed in 1972 by Isaac Nassi and the at the time graduate student Ben Shneiderman. These diagrams are also called structograms, as they show a program's structures.

Chapter 10. Entity Relationship Modeling

Entity-relationship modeling (ERM) is a conceptual modeling technique used primarily for software system representation. Entity-relationship diagrams, which are a product of executing the ERM technique, are normally used to represent database models and information systems. The main components of the diagram are the entities and relationships. The entities can represent independent functions, objects, or events. The relationships are responsible for relating the entities to one another. To form a system process, the relationships are combined with the entities and any attributes needed to describe further the process.

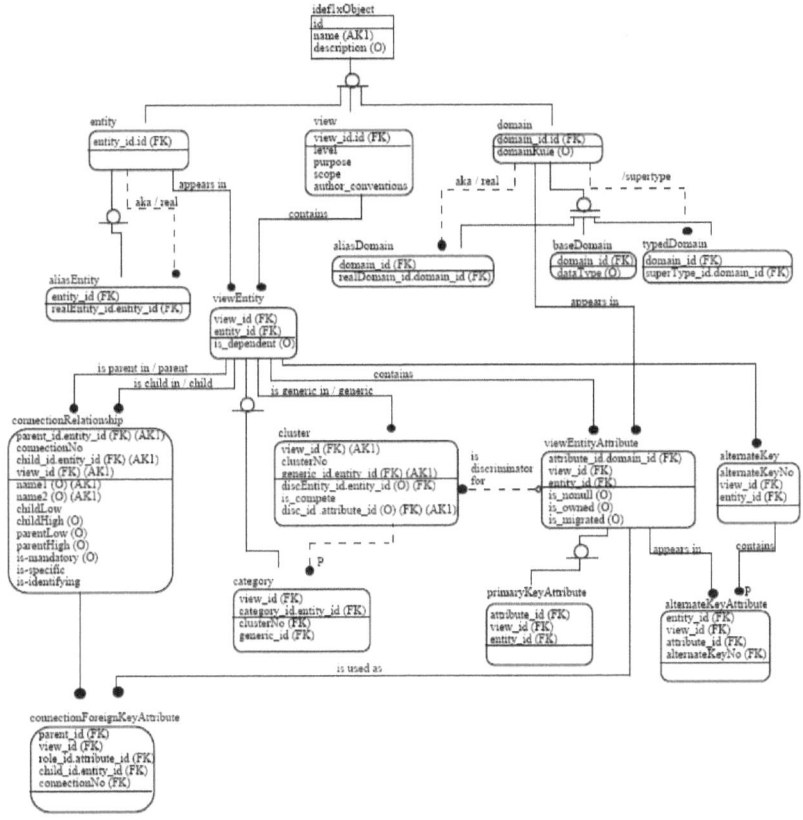

Figure 10-1. Example of an IDEF1X Entity relationship diagrams

Multiple diagramming conventions exist for this technique: IDEF1X, Bachman, and EXPRESS, to name a few. These conventions are just different ways of viewing and organizing the data to represent different system aspects.

Entity-relationship model

In software engineering, an **entity-relationship model (ERM)** is an abstract and conceptual representation of data. Entity-relationship modeling is a database modeling method, used to produce a type of conceptual schema or semantic data model of a system, often a relational database, and its requirements in a top-down fashion. Diagrams created by this process are called **entity-relationship diagrams**, **ER diagrams**, or **ERDs**.

This chapter refers to the techniques proposed in Peter Pin-Shan Chen's[1] 1976 paper [83]. However, variants of the idea existed previously [84], and have been devised subsequently.

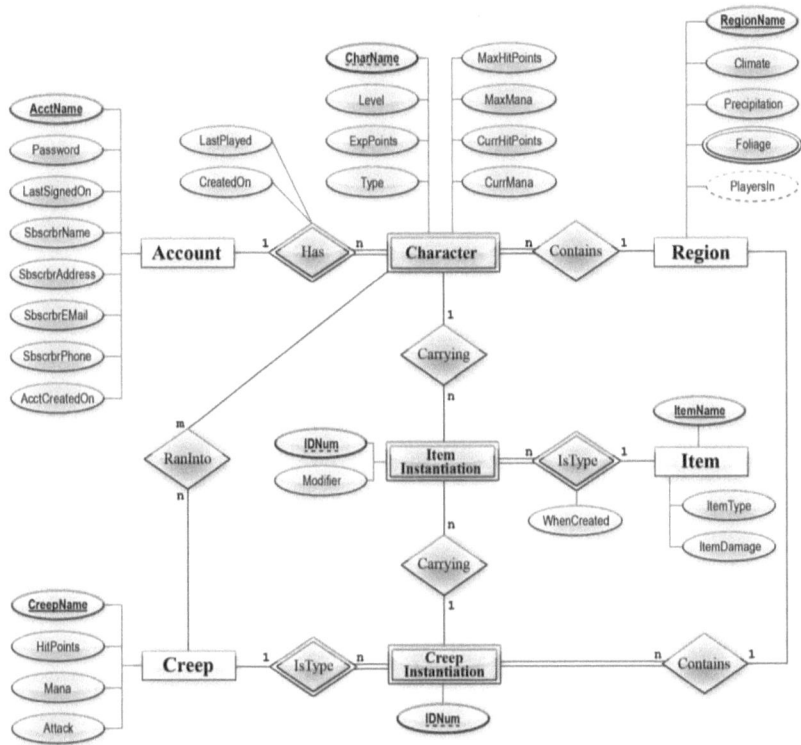

Figure 10-2. A sample Entity-relationship diagram using Chen's notation

Overview

The first stage of information system design uses these models during the requirements analysis to describe information needs or the type of information that is to be stored in a database. The data modeling technique can be used to describe any ontology (i.e. an overview and classifications of used terms and their relationships) for a certain area of interest. In the case of the design of an information system that is based on a database, the conceptual data model is, at a later stage (usually called logical design), mapped to a logical data model[2], such as the relational model[3]; this in turn is mapped to a physical model during physical design. Note that sometimes, both of these phases are referred to as "physical design".

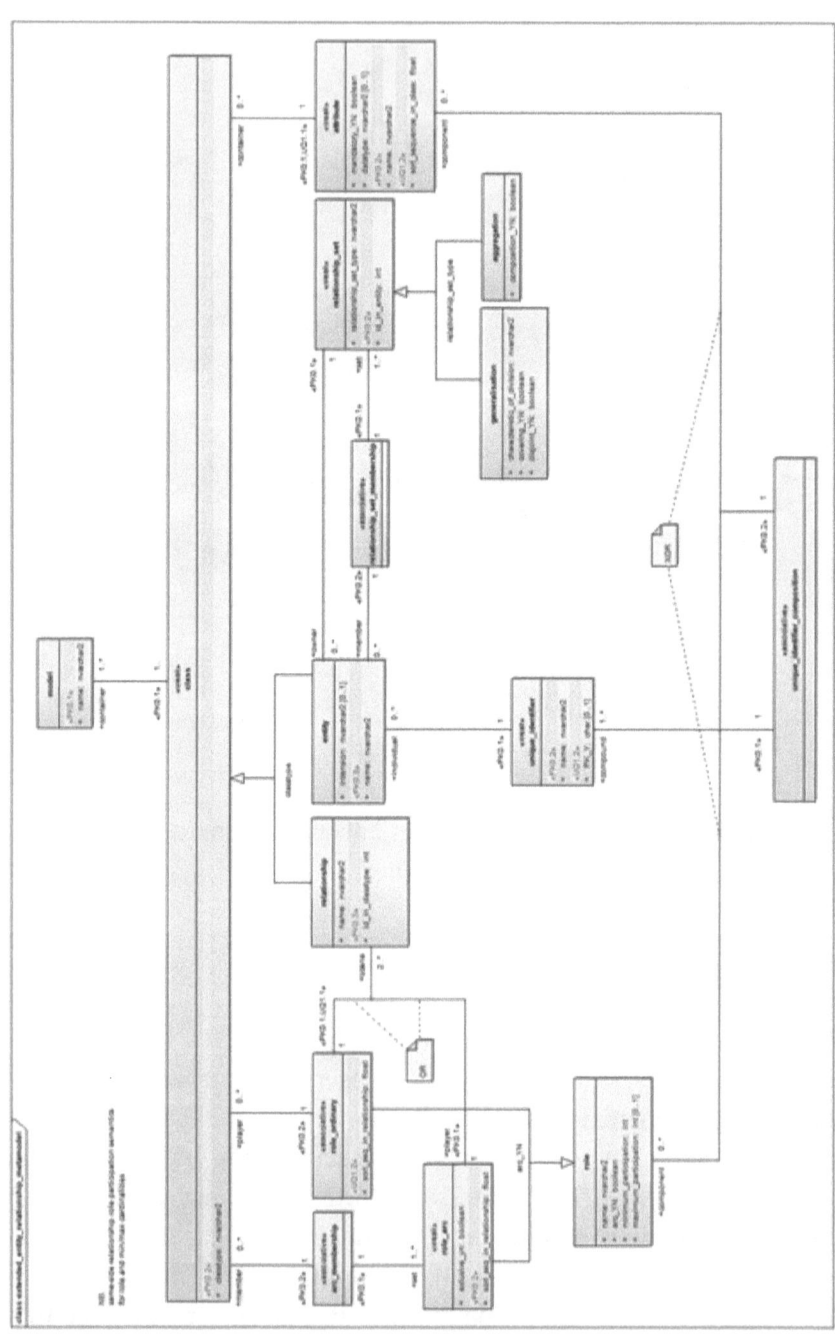

Figure 10-3. A UML metamodel of Extended Entity Relationship models

The building blocks: entities, relationships, and attributes

Figure 10-4(a). Two related entities

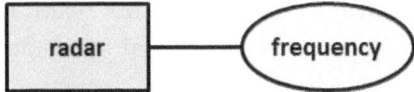

Figure 10-4(b). An entity with an attribute

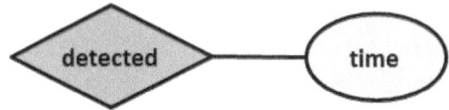

Figure 10-4(c). A relationship with an attribute

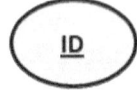

Figure 10-4(d). Primary key

We may define an entity as a thing which is recognized as being capable of an independent existence and which can be uniquely identified. An entity is an abstraction from the complexities of some domain. When we speak of an entity, we normally speak of some aspect of the real world that can be distinguished from other aspects of the real world [85].

An entity may be a physical object such as a house or a car, an event such as a house sale or a car service, or a concept such as a customer transaction or order. Although the term entity is the one most commonly used, following Chen [83] we should really distinguish between an entity and an entity-type. An entity-type is a category. An entity, strictly speaking, is an instance of a given entity-type. There are usually many instances of an entity-type. Because the term entity-type is somewhat cumbersome, most people tend to use the term entity as a synonym for this term. Entities can be thought of as nouns. Examples include a computer, an employee, a missile, a radar, a mathematical theorem.

A relationship captures how two or more entities are related to one another. Relationships can be thought of as verbs, linking two or more nouns. Examples include an *'owns'* relationship between a company and a computer, a *'supervises'* relationship between an employee and a department, a *'fires'* relationship between an launcher and a missile, a *'proved'* relationship between a mathematician and a theorem.

The model's linguistic aspect described above is utilized in the declarative database query language ERROL, which mimics natural language constructs. ERROL's semantics and implementation are based on Reshaped relational algebra (RRA), a relational algebra, which is adapted to the ERM and captures its linguistic aspect.

Entities and relationships can both have attributes. Examples include an *employee* entity might have a *Social Security Number* (SSN) attribute; the *radar* relationship may have a *frequency* attribute.

Every entity (unless it is a weak entity) must have a minimal set of uniquely identifying attributes, which is called the entity's primary key.

Entity-relationship diagrams do not show single entities or single instances of relations. Rather, they show entity sets and relationship sets.

> Example: a particular *missile* is an entity. The collection of all missiles in a database is an entity set. The *tracked* relationship between a seeker and a tactical ballistic missile (TBM) is a single relationship. The set of all such seeker-TBM relationships in a database is a relationship set. In other words, a relationship set corresponds to a relation in mathematics, while a relationship corresponds to a member of the relation.

Certain cardinality constraints[4] on relationship sets may be indicated as well.

Relationships, roles and cardinalities

In Chen's original paper [83], he gives an example of a relationship and its roles. He describes a relationship "marriage" and its two roles "husband" and "wife".

A person plays the role of husband in a marriage (relationship) and another person plays the role of wife in the (same) marriage. These words are nouns. That is no surprise, naming things requires a noun.

However, as is quite usual with new ideas, many eagerly appropriated the new terminology but then applied it to their own old ideas. Thus, the lines, arrows and crows-feet of their diagrams owed more to the earlier Bachman diagrams[5] than to Chen's relationship diamonds. Moreover, they similarly misunderstood other important concepts.

In particular, it became fashionable (now almost to the point of exclusivity) to "name" relationships and roles as verbs or phrases.

Relationship Names

A relationship expressed with a verb that implies direction, makes it impossible to discuss the model using correct English. Examples are:

- the song and the performer are related by a 'performs'
- the husband and wife are related by an 'is-married-to'.

Expressing the relationships with a noun resolves this:

- the song and the performer are related by a 'performance'
- the husband and wife are related by a 'marriage'.

Role naming

It has also become prevalent to name roles with phrases e.g. is-the-owner-of and is-owned-by etc. Correct nouns in this case are "owner" and "possession". Thus "person plays the role of owner" and "car plays the role of possession" rather than "person plays the role of is-the-owner-of" etc.

The use of nouns has direct benefit when generating physical implementations from semantic models. When a person has two relationships with car then it is possible to generate very simply names such as "owner_person" and "driver_person" which are immediately meaningful.

Cardinalities

However, some modifications to the original specification are beneficial. Chen described look-across cardinalities. UML perpetuates this. (As an aside, the Barker-Ellis notation used in *Oracle Designer* employs same-side for minimum cardinality (analogous to optionality) and role, but look-across for maximum cardinality (the crow's foot)).

Other authors (Merise [86], Elmasri & Navathe [87] amongst others [88]) prefer same-side for roles and both minimum and maximum cardinalities. Recent researchers (Feinerer [89], Dullea et. alia [90]) have shown that this is more coherent when applied to n-ary relationships of order > 2.

In Dullea et.al., *"An analysis of structural validity in entity-relationship modeling"*, one reads *"A 'look across' notation such as used in the UML does not effectively represent the semantics of participation constraints imposed on relationships where the degree is higher than binary."*

In Ingo Feinerer [89], we read, *"Problems arise if we operate under the look-across semantics as used for UML associations. Hartmann investigates this situation and shows how and why different transformations fail."* [91] (Although the "reduction" mentioned is spurious as the two diagrams 3.4 and 3.5 are in fact the same) and also *"As we will see on the next few pages, the look-across interpretation introduces several difficulties which prevent the extension of simple mechanisms from binary to n-ary associations."*

Semantic Modeling

The father of ER modeling said in his seminal paper: *"The entity-relationship model adopts the more natural view that the real world consists of entities and relationships. It incorporates some of the important semantic information about the real world"* [92]. He is here in accord with philosophic and theoretical traditions from the time of the Ancient Greek philosophers: Socrates, Plato and Aristotle (428 BC) through to modern epistemology, semiotics[6] and logic of Pierce, Frege and Russell. Plato himself associates knowledge with the apprehension of unchanging Forms (The forms, according to Socrates, are roughly speaking archetypes or abstract representations of the many types of things, and properties) and their relationships to one another. In his original 1976 article, Chen explicitly contrasts Entity-Relationship diagrams with record modeling techniques: *"The data structure diagram is a representation of the organization of records and is not an exact representation of entities and relationships"* [92]. Several other authors also support his program:

William Kent in "Data and Reality" [93]: *"One thing we ought to have clear in our minds at the outset of a modeling endeavor is whether we are intent on describing a portion of 'reality' (some human enterprise) or a data processing activity."*

Jean-Raymond Abrial[7] in "Data Semantics" [94]: *"... the so called 'logical' definition and manipulation of data are still influenced (sometimes unconsciously) by the 'physical' storage and retrieval mechanisms currently available on computer systems."*

Ronald Stamper[8] [95]: *"They pretend to describe entity types, but the vocabulary is from data processing: fields, data items, values. Naming rules don't reflect the conventions we use for naming people and things; they reflect instead techniques for locating records in files."*

In Michael A. Jackson's[9] words [96]: *"The developer begins by creating a model of the reality with which the system is concerned, the reality which furnishes its [the system's] subject matter ..."*

Elmasri and Navathe[10] [97]: *"The ER model concepts are designed to be closer to the user's perception of data and are not meant to describe the way in which data will be stored in the computer."*

A semantic model is a model of concepts; it is sometimes called a "platform independent model". It is an intentional model. At the least since Rudolf Carnap[11], it is well known that [98]: *"...the full meaning of a concept is constituted by two aspects, its intension and its extension. The first part comprises the embedding of a concept in the world of concepts as a whole, i.e. the totality of all relations to other concepts. The second part establishes the referential meaning of the concept, i.e. its counterpart in the real or in a possible world"*. An extensional model is that which maps to the elements of a particular methodology or technology, and is thus a "platform specific model". The UML specification explicitly states that associations in class models are extensional and this is in fact self evident by considering the extensive array of additional "adornments" provided by the specification over and above those provided by any of the prior candidate "semantic modeling languages" [99].

Diagramming conventions

Chen's notation for entity-relationship modeling uses rectangles to represent entities, and diamonds to represent relationships appropriate for first-class objects[12]: they can have attributes and relationships of their own. Entity sets are drawn as rectangles, relationship sets as diamonds. If an entity set participates in a relationship set, they are connected with a line.

Attributes are drawn as ovals and are connected with a line to exactly one entity or relationship set.

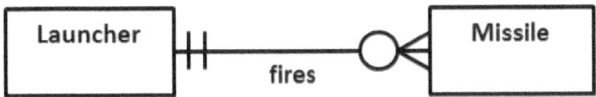

Figure 10-5. Two related entities shown using Crow's Foot notation. In this example, an optional relationship is shown between Artist and Song; the symbols closest to the song entity represents "zero, one, or many", whereas a song has "one and only one" Artist. Therefore, the former is read as, a Launcher (can) fire(s) "zero, one, or many" missile(s).

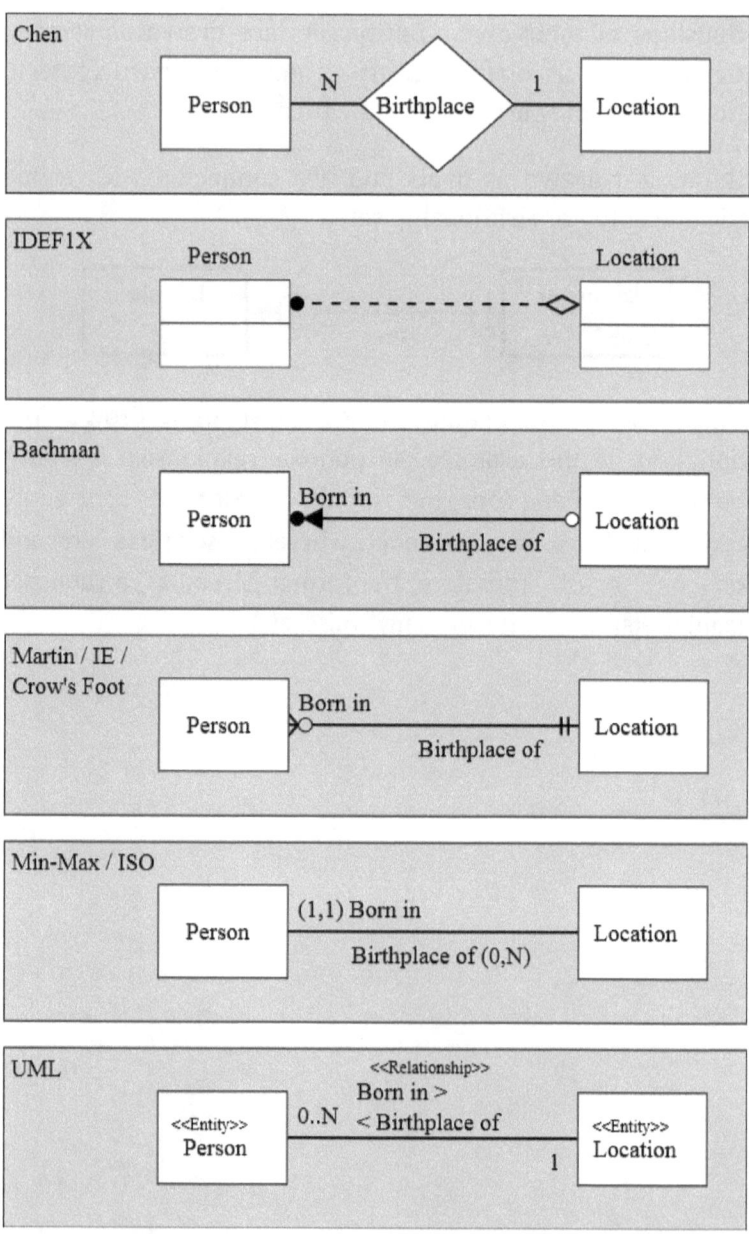

Figure 10-6. Various methods of representing the same one to many relationships. In each case, the diagram shows the relationship between a person and a place of birth: each person must have been born at one, and only one, location, but each location may have had zero or more people born at it.

Cardinality constraints are expressed as follows:

- a double line indicates a *participation constraint*, totality or surjectivity: all entities in the entity set must participate in *at least one* relationship in the relationship set;
- an arrow from entity set to relationship set indicates a key constraint, i.e. injectivity: each entity of the entity set can participate in *at most one* relationship in the relationship set;
- a thick line indicates both, i.e. bijectivity: each entity in the entity set is involved in *exactly one* relationship.
- an underlined name of an attribute indicates that it is a key: two different entities or relationships with this attribute always have different values for this attribute.

We often omit attributes as they can clutter up a diagram; other diagram techniques often list entity attributes within the rectangles drawn for entity sets.

Related diagramming convention techniques:

- Bachman notation
- EXPRESS[13]
- IDEF1X[14] [100]
- Martin notation[15]
- (min, max)-notation[16] of Jean-Raymond Abrial in 1974
- UML class diagrams[17]
- Merise[18]

Crow's Foot Notation

Crow's Foot notation is used in Barker's Notation[19], SSADM and Information Engineering. Crow's Foot diagrams represent entities as boxes, and relationships as lines between the boxes. Different shapes at the ends of these lines represent the cardinality of the relationship.

Crow's Foot notation was used in the 1980s by the consultancy practice CACI. Many of the consultants at CACI (including Richard Barker) subsequently moved to Oracle UK, where they developed the early versions of Oracle's CASE tools, introducing the notation to a wider audience. The following tools use Crow's Foot notation: ARIS, System Architect, Visio, PowerDesigner, Toad Data Modeler, DeZign for Databases, Devgems Data Modeler, OmniGraffle, MySQL Workbench and SQL Developer Data Modeler. CA's ICASE tool, CA Gen aka Information Engineering Facility also uses this notation.

ER diagramming tools

There are many ER diagramming tools. Some free software ER diagramming tools that can interpret and generate ER models and SQL and do database analysis are MySQL Workbench (formerly DBDesigner), and Open ModelSphere (open-source). A freeware ER tool that can generate database and application layer code (webservices) is the RISE Editor. The Open Source Erviz takes a simple textual description of the ERD and then uses Graphviz to produce automatically a layout. The web application TinyModeler allows you to make ER diagrams in a web browser.

Some of the proprietary ER diagramming tools are ARIS, Avolution, Aqua Data Studio, dbForge Studio for MySQL, DeZign for Databases, ER/Studio, Devgems Data Modeler, ERwin, MEGA International, ModelRight, OmniGraffle, Oracle Designer, Oracle Data Modeler, PowerDesigner, Rational Rose, Sparx Enterprise Architect, SQLyog, System Architect, Toad Data Modeler, SQL Maestro, Microsoft Visio, Visible Analyst, and Visual Paradigm.

Some free software diagram tools just draw the shapes without having any knowledge of what they mean, nor do they generate SQL. These include yEd, LucidChart, Gliffy [101], Kivio, and Dia. DIA diagrams, however, can be translated with tedia2sql.

Notes

[1] Dr. Peter Pin-Shan Chen is an American computer scientist and Professor of Computer Science at Louisiana State University, who is known for the development of Entity-Relationship Modeling in 1976. The Entity-Relationship Model serves as the foundation of many systems analysis and design methodologies, computer-aided software engineering (CASE) tools, and repository systems.

[2] A logical data model (LDM) in systems engineering is a representation of an organization's data, organized in terms of entities and relationships and is independent of any particular data management technology. Logical data models represent the abstract structure of some domain of information. They are often diagrammatic in nature and are most typically used in business processes that seek to capture things of importance to an organization and how they relate to one another. Once validated and approved, the logical data model can become the basis of a physical data model and inform the design of a database.

[3] The relational model for database management is a database model based on first-order predicate logic, first formulated and proposed in 1969 by Edgar F. Codd. The purpose of the relational model is to provide a declarative method for specifying data and queries: users directly state what information the database contains and what information they want from it, and let the database management system software take care of describing data structures for storing the data and retrieval procedures for answering queries.

[4] In data modeling, the cardinality of one data table with respect to another data table is a critical aspect of database design. Relationships between data tables define cardinality when explaining how each table links to another. In the relational model, tables can be related as any of: many-to-many, many-to-one (rev. one-to-many), or one-to-one. This is said to be the cardinality of a given table in relation to another.

[5] Bachman diagrams are diagrams which are used to design the data using a network or relational "logical" model, separating the data model from the way the data is stored in the system. The model is named after database pioneer Charles Bachman, and mostly used in computer software design. A Bachman diagram is another name for a data structure diagram

⁶ Semiotics, also called semiotic studies or (in the Saussurean tradition) semiology, is the study of signs and sign processes (semiosis), indication, designation, likeness, analogy, metaphor, symbolism, signification, and communication. Semiotics is closely related to the field of linguistics, which, for its part, studies the structure and meaning of language more specifically. Semiotics is often divided into three branches: Semantics: Relation between signs and the things to which they refer; their denotata, or meaning; Syntactics: Relations among signs in formal structures; Pragmatics: Relation between signs and the effects they have on the people who use them

⁷ Jean-Raymond Abrial is the father of the Z notation (typically used for formal specification of software), during his time at the Programming Research Group within the Oxford University Computing Laboratory, and later the B-Method (normally used for software development), two leading formal methods for software engineering. He is the author of The B-Book: Assigning Programs to Meanings (ISBN 0-521-49619-5). For much of his career he has been an independent consultant, as much at home working with industry as academia. Latterly, he became a Professor at ETH Zurich in Switzerland.

⁸ Ronald Stamper (born 1934 in West Bridgford, United Kingdom) is a British computer scientist, formerly a researcher in the London School of Economics and Professor at the University of Twente, known for his pioneering work in Organizational semiotics, and the creation of the MEASUR methodology and the SEDITA framework.

⁹ Michael Anthony Jackson (born 1936) is a British computer scientist, and independent computing consultant in London, England. He is also part-time researcher at AT&T Research, Florham Park, NJ, U.S., and visiting research professor at the Open University in the UK.

¹⁰ Shamkant B. Navathe is a noted researcher in the field of databases with more than 150 publications on different topics in the area of databases. He is presently a Professor in the College of Computing at Georgia Institute of Technology and founded the Research Group in Database Systems at the College of Computing at Georgia Institute of Technology.

¹¹ Rudolf Carnap (May 18, 1891 – September 14, 1970) was an influential German-born philosopher who was active in Europe before 1935 and in the United States thereafter. He was a major member of the Vienna Circle and an advocate of logical positivism.

[12] In programming language design, a first-class citizen (also object, entity, or value), in the context of a particular programming language, is an entity that can be constructed at run-time, passed as a parameter, returned from a subroutine, or assigned into a variable.

[13] EXPRESS is a standard data modeling language for product data. EXPRESS is formalized in the ISO Standard for the Exchange of Product model STEP (ISO 10303), and standardized as ISO 10303-11.

[14] IDEF1X (Integration Definition for Information Modeling) is a data modeling language for the developing of semantic data models. IDEF1X is used to produce a graphical information model, which represents the structure and semantics of information within an environment or system.

[15] The Martin notation (also Crow's foot notation; also known as crow's foot notation) due to James Martin, Bachmann and Odell is a notation for the semantic data modeling to represent simplified entity-relationship models.

[16] The min-max notation (also (min, max) notation) is a way to restrict the cardinality of a relationship between entity types in an entity relationship model. It was introduced because the Chen notation was allowed only limited statements to a relation. With the (min, max) notation can lower as also upper limits are expressed in accurately.

[17] In software engineering, a class diagram in the Unified Modeling Language (UML) is a type of static structure diagram that describes the structure of a system by showing the system's classes, their attributes, operations (or methods), and the relationships among the classes.

[18] Merise is a general-purpose modeling methodology in the field of information systems development, software engineering and project management. First introduced in the early 1980s, it was widely used in France, and was developed and refined to the point where most large French governmental, commercial and industrial organizations had adopted it as their standard methodology.

[19] Barker's notation refers to the ERD notation developed by Richard Barker, Ian Palmer, Harry Ellis et al. while working at the British consulting firm CACI around 1981. The notation was adopted by Barker when he joined Oracle and is effectively defined in his book Entity Relationship Modeling as part of the CASE Method series of books. This notation was and still is used by the Oracle CASE modeling tools. It is a variation of the

"crows foot" style of data modeling that was favored by many over the original Chen style of ERD modeling because of its readability and efficient use of drawing space.

Chapter 11. Event-Driven Process Chain

The event-driven process chain (EPC) is a conceptual modeling technique that is mainly used to improve systematically business process flows. Like most conceptual modeling techniques, the event driven process chain consists of entities/elements and functions that allow relationships to be developed and processed. More specifically, the EPC is made up of events that define what state a process is in or the rules by which it operates. In order to progress through events, a function/ active event must be executed. Depending on the process flow, the function has the ability to transform event states or link to other event driven process chains. Other elements exist within an EPC, all of which work together to define how and by what rules the system operates. The EPC technique can be applied to business practices such as resource planning, process improvement, and logistics.

Event-driven process chain

An **Event-driven Process Chain** (EPC) is a type of flowchart used for business process modeling[1]. EPC's can be used for configuring an enterprise resource planning (ERP) implementation[2] [102], and for business process improvement.

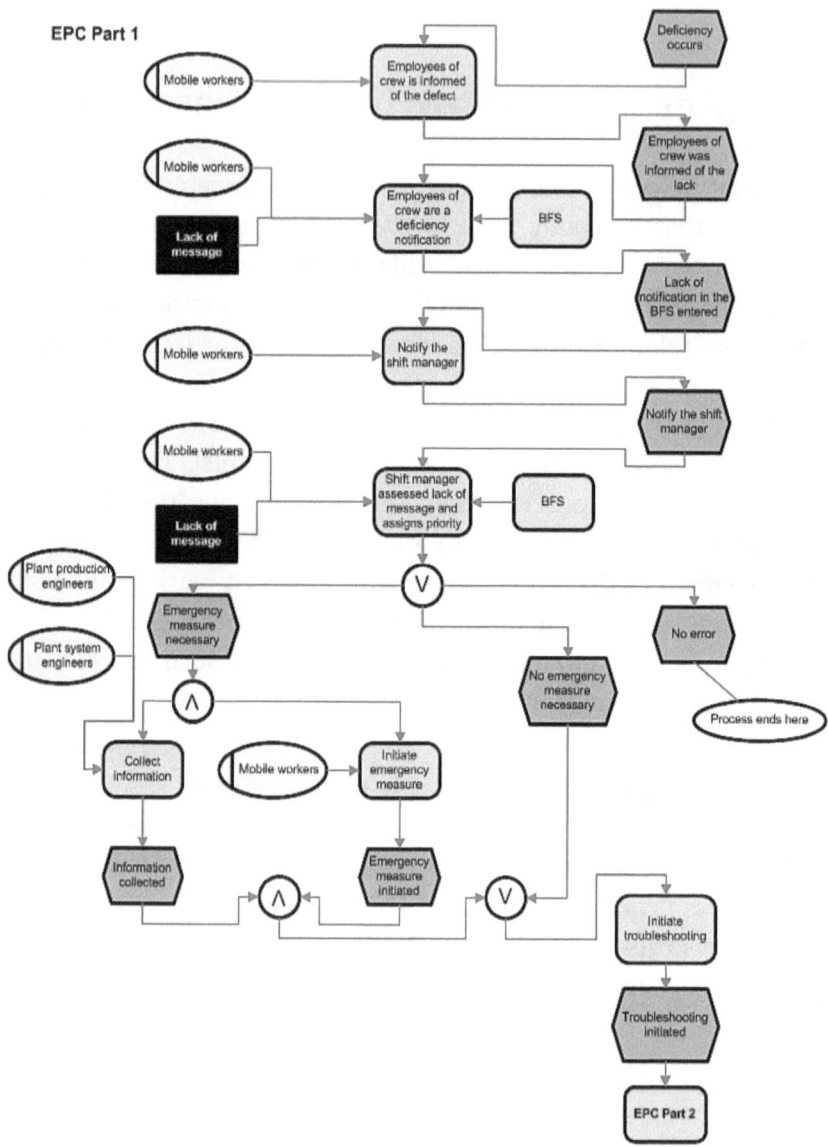

Figure 11-1. Example of a more complex EPC diagram.

Overview

Businesses use EPC diagrams to lay out business process workflows, originally in conjunction with SAP R/3 modeling[3], but now more widely. There are a number of tools for creating EPC diagrams: ARIS Toolset of IDS Scheer AG [103] (Now taken over by Software AG), free modeling tool ARIS Express by IDS Scheer AG, ADONIS of BOC Group [104], Mavim Rules of Mavim BV [105], Business Process Visual ARCHITECT of Visual Paradigm, Visio of Microsoft Corp., Semtalk of Semtation GmbH, or Bonapart by Pikos GmbH. Some but not all of these tools support the tool-independent EPC Markup Language (EPML) interchange format. There are also tools that generate EPC diagrams from operational data, such as SAP logs. EPC diagrams use symbols of several kinds to show the control flow structure (sequence of decisions, functions, events, and other elements) of a business process.

The EPC method was developed within the framework of ARIS[4] by Prof. Wilhelm-August Scheer at the Institut für Wirtschaftsinformatik at the Universität des Saarlandes in the early 1990s [106]. Many companies use it for modeling, analyzing, and redesigning business processes. As such, it forms the core technique for modeling in ARIS, which serves to link the different views in the so-called control view, which we will elaborate in section of ARIS Business Process Modeling.

To quote from a publication on EPCs [107]: "An EPC is an ordered graph[5] of events and functions. It provides various connectors that allow alternative and parallel execution of processes. Furthermore it is specified by the usages of logical operators, such as OR, AND, and XOR. A major strength of EPC is claimed to be its simplicity and easy-to-understand notation. This makes EPC a widely acceptable technique to denote business processes."

The statement that EPCs are ordered graphs is also found in other literature, but is probably a misformulation: an ordered graph is an undirected graph with an explicitly provided total node ordering, while EPCs are directed graphs[6] for which no explicit node ordering is provided. No restrictions actually appear to exist on the possible structure of EPCs, but nontrivial structures involving parallelism have ill-defined execution semantics; in this respect, they resemble UML activity diagrams[7]. Several scientific articles are devoted to providing well-defined execution semantics for general EPCs [108] [109]. One particular issue is that EPCs require non-local semantics [110], i.e., the execution behavior of a particular node within an EPC may depend on the state of other parts of the EPC, arbitrarily far away.

In the following, we will describe the elements used in EPC diagram.

Elements of an event-driven process chain

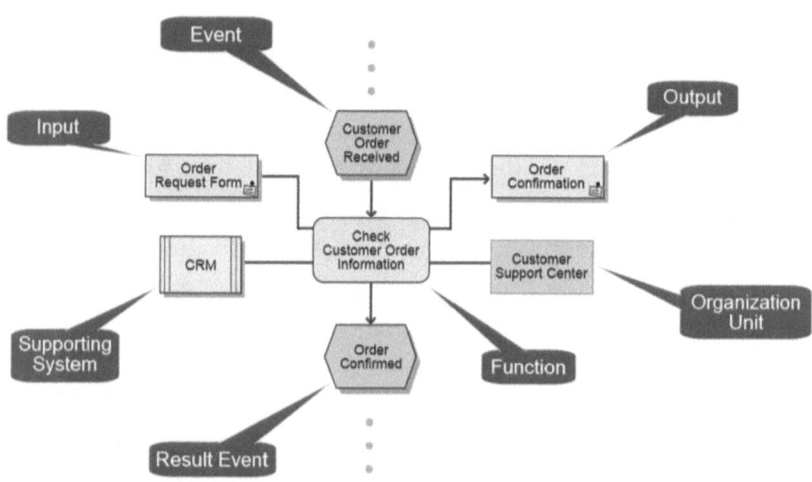

Figure 11-2. Elements of an event driven process

Event
Events are passive elements in EPC. They describe under what circumstances a function or a process works, or which state a function or a process results in. Examples of events are "requirement captured", "material on stock", etc. In the EPC graph,

an event is represented as hexagon. In general, an EPC diagram must start with an event and end with an event.

Function

Functions are active elements in EPC. They model the tasks or activities within the company. Functions describe transformations from an initial state to a resulting state. In case different resulting states can occur, the selection of the respective resulting state can be modeled explicitly as a decision function using logical connectors. Functions can be refined into another EPC. In this case, it is called hierarchical function. Examples of functions are "capture requirement", "check material on stock", etc. In the EPC graph, a function is represented as rounded rectangle.

Organization unit

Organization units determine which person or organization within the structure of an enterprise is responsible for a specific function. Examples are "sales department", "sales manager", "procurement manager", etc. It is represented as an ellipse with a vertical line.

Information, material, or resource object

In the EPC, the information, material, or resource objects portray objects in the real world, for example business objects, entities, etc., which can be input data serving as the basis for a function, or output data produced by a function. Examples are "material", "order", etc. In the EPC graph, such an object is represented as rectangle.

Logical connector

In the EPC the logical relationships between elements in the control flow, that is, events and functions are described by logical connectors. With the help of logical connectors, it is possible to split the control flow from one flow to two or more flows and to synchronize the control flow from two or more flows to one flow.

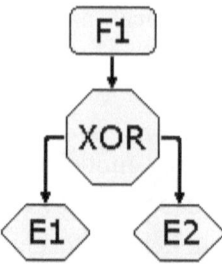 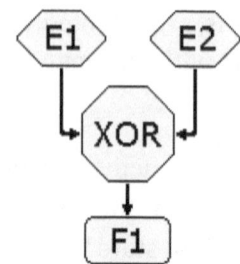

Figure 11-3(a). If function F1 completes, either events E1 or E2 occur

Figure 11-3(b). If either events E1 or E2 occur, function F1 starts

Logical relationships

There are three kinds of logical relationships defined in EPC:

- Branch/Merge: Branch and merge correspond to making decision of which path to choose among several control flows. A branch may have one incoming control flow and two or more outgoing control flows. When the condition is fulfilled, a branch activates exactly only one of the outgoing control flows and deactivates the others. The counterpart of a branch is a merge. A merge may have two or more incoming flows and one outgoing control flow. A merge synchronizes an activated and the deactivated alternatives. The control will then be passed to the next element after the merge. A branch in the EPC is represented by an opening XOR[8], whereas a merge is represented as a closing XOR connectors.
- Fork/Join: Fork and join correspond to activating all paths in the control flow concurrently. A fork may have one incoming control flow and two or more outgoing control flows. When the condition is fulfilled, a fork activates all of the outgoing control flows in parallel. A join may have two or more incoming control flows and one outgoing control flow. A join synchronizes all activated incoming control flows. In the EPC diagram how the

concurrency achieved is not a matter. In reality, the concurrency can be achieved by true parallelism or by virtual concurrency achieved by interleaving. A fork in the EPC is represented by an opening 'AND', whereas a join is represented as a closing 'AND' connectors.

- OR: An 'OR' relationship corresponds to activating one or more paths among control flows. An opening 'OR' connector may have one incoming control flow and two or more outgoing control flows. When the condition is fulfilled, an opening 'OR' connector activates one or more control flows and deactivates the rest of them. The counterpart of this is the closing 'OR' connector. When at least one of the incoming control flows is activated, the closing 'OR' connector will pass the control to the next element after it.

Control flow
A control flow connects events with functions, process paths, or logical connectors creating chronological sequence and logical interdependencies between them. A control flow is represented as a dashed arrow.

Information flow
Information flows show the connection between functions and input or output data, upon which the function reads changes or writes.

Organization unit assignment
Organization unit assignments show the connection between an organization unit and the function for which it is responsible.

Process path
Process paths serve as navigation aid in the EPC. They show the connection from or to other processes. The process path is represented as a compound symbol composed of a function symbol superimposed upon an event symbol. To employ the process path symbol in an EPC diagram, a symbol is connected to the process path symbol, indicating that the process diagramed incorporates the

entirety of a second process, which, for diagrammatic simplicity, is represented by a single symbol.

Example

As shown in the example, customer order received as the initial event, which creates a requirement capture within the company. In order to specify this function, sales is responsible for marketing, currency etc. As a result, event 'requirement captured' leads to another new function: check material on stock, in order to manufacture the productions. All input or output data about material remains in the information resource. After checking material, two events may happen-with or without material on stock. If positive, get material from stock; if not, order material from suppliers. Since the two situations cannot happen at the same time, XOR is the proper connector to link them together.

Meta-model of EPC

Although a real process may include a series of stages until it is finished eventually, the main activities remain similar. An event triggers one function; and a function will lead to one event. Meanwhile, an event may involve one or more processes to fulfill but a process is unique for one event, the same goes for Process and Process Path. As for the function, its data may be included in one or more information resources, while Organization Unit is only responsible for one specific function.

Notes

---

[1] Business Process Modeling (BPM) in systems engineering is the activity of representing processes of an enterprise, so that the current process may be analyzed and improved. BPM is typically performed by business analysts and managers who are seeking to improve process efficiency and quality. The process improvements identified by BPM may or may not

require Information Technology involvement, although that is a common driver for the need to model a business process, by creating a process master.

[2] Enterprise resource planning (ERP) integrates internal and external management information across an entire organization, embracing finance/accounting, manufacturing, sales and service, customer relationship management, etc. ERP systems automate this activity with an integrated software application. Its purpose is to facilitate the flow of information between all business functions inside the boundaries of the organization and manage the connections to outside stakeholders

[3] SAP R/3 is the former name of the main enterprise resource planning software produced by SAP AG. It is an enterprise-wide information system designed to coordinate all the resources, information, and activities needed to complete business processes such as order fulfillment or billing

[4] ARIS (Architecture of Integrated Information Systems) is an approach to enterprise modeling. It offers methods for analyzing processes and taking a holistic view of process design, management, workflow, and application processing.

[5] An ordered graph is a graph with a total order over its nodes. In an ordered graph, the parents of a node are the nodes that are joined to it and precede it in the ordering. The width of a node is the number of its parents, and the width of an ordered graph is the maximal width of its nodes.

[6] A directed graph is a graph whose edges have direction and are called arcs. Arrows on the arcs are used to encode the directional information: an arc from vertex A to vertex B indicates that one may move from A to B but not from B to A. A graph in which each graph edge is replaced by a directed graph edge, also called a digraph. A directed graph having no multiple edges or loops (corresponding to a binary adjacency matrix with 0s on the diagonal) is called a simple directed graph. A complete graph in which each edge is bidirected is called a complete directed graph. A directed graph having no symmetric pair of directed edges (i.e., no bidirected edges) is called an oriented graph. A complete oriented graph (i.e., a directed graph in which each pair of nodes is joined by a single edge having a unique direction) is called a tournament.

[7] Activity diagrams are graphical representations of workflows of stepwise activities and actions with support for choice, iteration and concurrency.

In the Unified Modeling Language, activity diagrams can be used to describe the business and operational step-by-step workflows of components in a system. An activity diagram shows the overall flow of control.

[8] A connective in logic known as the "exclusive or," or exclusive disjunction. It yields true if exactly one (but not both) of two conditions is true.

Chapter 12. Joint Application Development

The Dynamic Systems Development Method (DSDM) uses a specific process called Joint Application Design (JAD) to model conceptually a systems life cycle. JAD is intended to focus more on the higher level development planning that precedes a projects initialization. The JAD process calls for a series of workshops in which the participants work to identify, define, and generally map a successful project from conception to completion. This method has been found not to work well for large-scale applications, however smaller applications usually report some net gain in efficiency [17].

Joint application design

Joint application design (JAD) is a process used in the prototyping life cycle area of the Dynamic Systems Development Method (DSDM) to collect business requirements while developing new information systems for a company. *"The JAD process also includes approaches for enhancing user participation, expediting development, and improving the quality of specifications."* It consists of a workshop where *"knowledge workers and IT specialists meet, sometimes for several days, to define and review the business requirements for the system"* [111]. The attendees include high-level management officials who will ensure the product provides the needed reports and information at the end. This acts as *"a management process which allows Corporate Information Services (IS) departments to work more effectively with users in a shorter time frame"* [112].

Through JAD workshops the knowledge workers and IT specialists are able to resolve any difficulties or differences between the two parties regarding the new information system. The workshop follows a detailed agenda in order to guarantee that all uncertainties between parties are covered and to help prevent any miscommunications. Miscommunications can carry repercussions that are far more serious if not addressed until later on in the process. (See below for Key Participants and Key Steps to an Effective JAD). In the end, this process will result in a new information system that is feasible and appealing to both the designers and end users.

Although the JAD design is widely acclaimed, little is actually known about its effectiveness in practice. According to Journal of Systems and Software, a field study was done at three organizations using JAD practices to determine how JAD influenced system development outcomes. The results of the study suggest that organizations realized modest improvement in systems development outcomes by using the JAD method. JAD use was most effective in small, clearly focused projects and less effective in large complex projects.

Origin

Joint Application Development (JAD) Originally, JAD was designed to bring system developers and users of varying backgrounds and opinions together in a productive as well as creative environment. The meetings were a way of obtaining quality requirements and specifications. The structured approach provides a good alternative to traditional serial interviews by system analysts.

Key participants

Executive Sponsor: The executive who charters the project, the system owner. They must be high enough in the organization to be able to make decisions and provide the necessary strategy, planning, and direction.

Subject Matter Experts: These are the business users, the IS professionals, and the outside experts that will be needed for a successful workshop. This group is the backbone of the meeting; they will drive the changes.

Facilitator/Session Leader: Chairs the meeting and directs traffic by keeping the group on the meeting agenda. The facilitator is responsible for identifying those issues that can be solved as part of the meeting and those which need to be assigned at the end of the meeting for follow-up investigation and resolution. The facilitator serves the participants and does not contribute information to the meeting.

Scribe/Modeler/Recorder/Documentation Expert: Records and publish the proceedings of the meeting and does not contribute information to the meeting.

Observers: Generally, members of the application development team assigned to the project. They are to sit behind the participants and are to observe silently the proceedings.

9 Key Steps

Identify project objectives and limitations. It is vital to have clear objectives for the workshop and for the project as a whole. The pre-workshop activities, the planning and scoping, set the expectations of the workshop sponsors and participants. Scoping identifies the business functions that are within the scope of the project. It also tries to assess both the project design and implementation complexity. The political sensitivity of the project should be assessed. Has this been tried in the past? How many false starts were there? How many implementation failures were there? Sizing is important. For best results, systems projects should be sized so that a complete design - right down to screens and menus - can be designed in 8 to 10 workshop days.

Identify critical success factors. It is important to identify the critical success factors for both the development project and the business function being studied. How will we know that the planned changes have been effective? How will success be measured? Planning for outcomes assessment helps to judge the effectiveness and the quality of the implemented system over its entire operational life.

Define project deliverables. In general, the deliverables from a workshop are documentation and a design. It is important to define the form and level of detail of the workshop documentation. What types of diagrams will be provided? What type or form of narrative will be supplied? It is a good idea to start using a CASE tool for diagramming support right from the start. Most of the available tools have well to great diagramming capabilities but their narrative support is generally weak. The narrative is best produced with your standard word processing software.

Define the schedule of workshop activities. Workshops vary in length from one to five days. The initial workshop for a project should not be less than three days. It takes the participants most of the first day to get comfortable with their roles, with each other, and with the environment. The second day is spent learning to understand each other and developing a common language with which to communicate issues and concerns. By the third day, everyone is working together on the problem and real productivity is achieved. After the initial workshop, the team building has been done. Shorter workshops can be scheduled for subsequent phases of the project, for instance, to verify a prototype. However, it will take the participants from one to three hours to re-establish the team psychology of the initial workshop.

Select the participants. These are the business users, the IS professionals, and the outside experts that will be needed for a successful workshop. These are the true "back bones" of the meeting who will drive the changes.

Prepare the workshop material. Before the workshop, the project manager and the facilitator perform an analysis and build a preliminary design or straw man to focus the workshop. The workshop material consists of documentation, worksheets, diagrams, and even props that will help the participants understand the business function under investigation.

Organize workshop activities and exercises. The facilitator must design workshop exercises and activities to provide interim deliverables that build towards the final output of the workshop. The pre-workshop activities help design those workshop exercises. For example, for a Business Area Analysis, what is in it? A decomposition diagram? A high-level entity-relationship diagram? A normalized data model? A state transition diagram? A dependency diagram? All of the above? None of the above? It is important to define the level of technical diagramming that is appropriate to the environment. The most important thing about a diagram is that the user must understand it. Once the diagram choice is made, the facilitator designs exercises into the workshop agenda to get the group to develop those diagrams. A workshop combines exercises that are serially oriented to build on one another, and parallel exercises, with each sub-team working on a piece of the problem or working on the same thing for a different functional area. High-intensity exercises led by the facilitator energize the group and direct it towards a specific goal. Low-intensity exercises allow for detailed discussions before decisions. The discussions can involve the total group or teams can work out the issues and present a limited number of suggestions for the whole group to consider. To integrate the participants, the facilitator can match people with similar expertise from different departments. To help participants learn from each other, he can mix the expertise. It is up to the facilitator to mix and match the sub-team members to accomplish the organizational, cultural, and political objectives of the workshop. A workshop operates on both the technical level and the political level. It is the facilitator's job to build consensus and communications, to force issues out early in the process. There is no need to worry about the technical implementation of a system if the underlying business issues cannot be resolved.

Prepare, inform, and educate the workshop participants. All of the participants in the workshop must be made aware of the objectives and limitations of the project and the expected deliverables of the workshop. Briefing of participants should take place 1 to 5 days before the workshop. This briefing may be teleconferenced if participants are widely dispersed. The briefing document might be called the Familiarization Guide, Briefing Guide, Project Scope Definition, or the Management Definition Guide - or anything else that seems appropriate. It is a document of eight to twelve pages, and it provides a clear definition of the scope of the project for the participants. The briefing itself lasts two to four hours. It provides the psychological preparation everyone needs to move forward into the workshop.

Coordinate workshop logistics. Workshops should be held off-site to avoid interruptions. Projectors, screens, PCs, tables, markers, masking tape, Post-It notes, and lots of other props should be prepared. What specific facilities and props are needed is up to the facilitator. They can vary from simple flip charts to electronic white boards. In any case, the layout of the room must promote the communication and interaction of the participants.

Advantages

JAD decreases time and costs associated with requirements elicitation process. During 2-4 weeks, information not only is collected, but requirements, agreed upon by various system users, are identified. Experience with JAD allows companies to customize their systems analysis process into even more dynamic ones like Double Helix[1], a methodology for mission-critical work.

JAD sessions help bring experts together giving them a chance to share their views, understand views of others, and develop the sense of project ownership.

The methods of JAD implementation are well known, as it is "the first accelerated design technique available on the market and probably best known", and can easily be applied by any organization.

Easy integration of CASE tools into JAD workshops improves session productivity and provides systems analysts with discussed and ready to use models.

Challenges

1. "Do your homework". Without multifaceted preparation for a JAD session, valuable time of professionals can be wasted easily. The wrong problem can be addressed, the wrong people can be invited to participate, inadequate resources for problem solving can be used - all these scenarios can happen if organizers of the JAD session do not study the elements of the system being evaluated.

2. The team chosen to participate in a JAD workshop should include employees able to provide input on most, if not all, of the necessary parts of the problem. That is why particular attention should be paid during participant selection. The group should consist not only of employees from various departments who will interact with the new system, but also from different places on the organizational ladder. This variety of thought process understanding will reflect different, sometimes even conflicting points of view, but will allow participants to see a "different side of the coin". With better understanding of the undercurrents of processes JAD will bring to light a better model outline.

3. The facilitator as a smoothing and motivational force has to make sure that all participants, not the most vocal ones only, have a chance to offer their opinions, ideas, and thoughts. All business experts on the JAD team should be encouraged to offer their input, making discussions more fruitful.

Notes

[1] The Double Helix methodology is a systems development methodology used by Lockheed Martin. It combines experimentation, technology, and a warfighter's concept of operations to create new tactics and weapons.

Chapter 13. Dynamic systems development method

Dynamic systems development method (DSDM) is an agile project delivery framework, primarily used as a software development method. DSDM was originally based upon the rapid application development method. In 2007, DSDM became a generic approach to project management and solution delivery. DSDM is an iterative and incremental approach that embraces principles of Agile development[1], including continuous user/customer involvement.

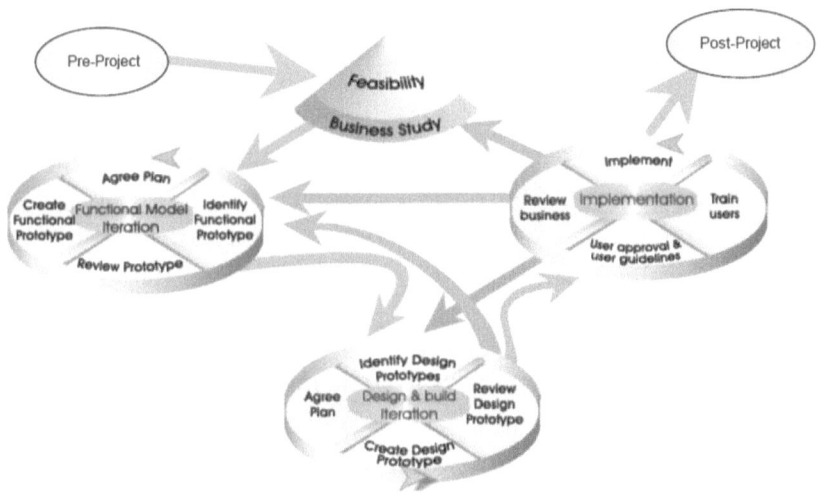

Figure 13-1. Project phases in SAM Atern [113]

DSDM fixes cost, quality and time at the outset and uses the MoSCoW prioritization[2] of scope into *musts, shoulds, coulds* and *won't haves* to adjust the project deliverable to meet the stated time constraint. DSDM is one of a number of Agile methods for developing software and non-IT solutions, and it forms a part of the Agile Alliance [114].

The most recent version of DSDM, launched in 2007, is called DSDM Atern. The name Atern is a shortening of Arctic Tern[3] - a collaborative bird that can travel vast distances and epitomizes many facets of the method that are natural ways of working e.g. prioritization and collaboration.

The previous version of DSDM (released in May 2003) which is still widely used and is still valid is DSDM 4.2, which is a slightly extended version of DSDM version 4. The extended version contains guidance on how to use DSDM with Extreme Programming[4].

DSDM and the DSDM Consortium: origins

In the early 1990s, a new term (RAD) was spreading across the IT industry. The user interfaces for software applications were moving from the old green screens to the graphical user interfaces that are used today. New application development tools were coming on the market, such as PowerBuilder. These enabled developers to share their proposed solutions much more easily with their customers – prototyping became a reality and the frustrations of the classical, sequential (waterfall) development methods could be put to one side.

However, the RAD movement was much unstructured: there was no commonly agreed definition of a suitable process and many organizations came up with their own definition and approach. Many major corporations were very interested in the possibilities but they were also concerned that they did not lose the level of quality in the end deliverables that free-flow development could give rise to.

The DSDM Consortium was founded in 1994 by an association of vendors and experts in the field of software engineering and was created with the objective of *"jointly developing and promoting an independent RAD framework"* [115] by combining their best practice experiences. The origins were an event organized by the Butler Group in London. People at that meeting all worked for blue-chip organizations such as British Airways, American Express, Oracle and Logica (other companies such as Data Sciences and Allied Domecq have since been absorbed by other organizations).

At the initial meeting, it was decided that Jennifer Stapleton, then of Logica, would put together an architecture for an end-to-end, user-centric but quality-controlled method for iterative and incremental development. The resulting architecture was designed to be fully compatible with ISO 9000 and PRINCE2, which were two major concerns for the group. Once the architecture was in place (a month after the initial meeting), the Consortium formed various task groups to populate it with all aspects of software development, including project management tools and techniques, quality and testing, development tools and techniques, personnel and software procurement. An oversight group led by the architect and consisting of the chairs of the task groups, ensured consistency of the approach as it was developed.

Although many of the Consortium members were direct business competitors, they shared freely how they had addressed the various aspects. Best practice was extracted and formed into a cohesive whole. As the Consortium grew in its first year from a handful of organizations to sixty, the content of the method became increasingly robust. Version 1 was baselined in December 1994 and published in February 1995. The result was a generic method covering people, process and tools that was formed from the experiences of organizations of all sectors and sizes.

The DSDM Consortium is a not-for-profit, vendor-independent organization which owns and administers the DSDM framework [115].

DSDM Atern

Atern is a vendor-independent approach that recognizes that more projects fail because of people issues than technology. Atern's focus is on helping people to work effectively together to achieve the business goals. Atern is also independent of tools and techniques enabling it to be used in any business and technical environment without tying the business to a particular vendor [116].

Overview of DSDM Atern

As an extension of rapid application development, DSDM focuses on information systems projects that are characterized by tight schedules and budgets. DSDM addresses the most common failures of information systems projects, including exceeding budgets, missing deadlines, and lack of user involvement and top-management commitment

In 2007, a team set up by the DSDM Consortium looked into the content of DSDM V4.2 and decided that the underlying mechanics and structure were completely sound but that the terminology and the focus purely on IT applications should be updated to meet the needs of projects in the new millennium. Some of the content of the method had been there since 1995. The new version was launched at the Cafe Royale in London on 24 April 2007. Some amendments were made in April 2008 (Atern V2) and incorporated in the latest version of the Atern Handbook [117].

The DSDM Atern approach
Principles

There are eight principles underpinning DSDM Atern. These principles direct the team in the attitude they must take and the mindset they must adopt in order to deliver consistently.

1. Focus on the business need
The main criteria for acceptance of a "deliverable" is delivering a system that addresses the current business needs. Delivering a perfect system that addresses all possible business needs is less important than focusing on critical functionalities.

- Understand the true business priorities
- Establish a sound Business Case

- Seek continuous business sponsorship and commitment
- Guarantee the Minimum Usable Subset of features.

2. Deliver on time
- Timebox[5] the work
- Focus on business priorities
- Always hit deadlines

3. Collaborate
User involvement is the main key in running an efficient and effective project, where both users and developers share a workplace (either physical or via tools), so that the decisions can be made collaboratively and quickly.

- Involve the right stakeholders, at the right time, throughout the project
- Ensure that the members of the team are empowered to take decisions on behalf of those they represent without waiting for higher-level approval.
- Actively involve the business representatives
- Build a one-team culture

4. Never compromise quality
- Set the level of quality at the outset
- Ensure that quality does not become a variable
- Design, document and test appropriately
- Build in quality by constant review
- Test early and continuously. See test-driven development[6] for comparison.

5. Build incrementally from firm foundations
- Strive for early delivery of business benefit where possible
- Continually confirm the correct solution is being built
- Formally re-assess priorities and ongoing project viability with each delivered increment

6. Develop iteratively

A focus on frequent delivery of products, with assumption that to deliver something "good enough" earlier is always better than to deliver everything "perfectly" in the end. By delivering product frequently from an early stage of the project, the product can be tested and reviewed where the test record and review document can be taken into account at the next iteration or phase.

- Do enough design up front to create strong foundations
- Take an iterative approach to building all products
- Build customer feedback into each iteration to converge on an effective business solution
- Accept that most detail emerges later rather than sooner
- Embrace change – the right solution will not evolve without it
- Be creative, experiment, learn, evolve

7. Communicate continuously and clearly

Communication and cooperation among all project stakeholders is required to be efficient and effective.

- Run daily team stand-up sessions
- Use facilitated workshops
- Use rich communication techniques such as modeling and prototyping
- Present iterations of the evolving solution early and often
- Keep documentation lean and timely
- Manage stakeholder expectations throughout the project
- Encourage informal, face to face communication at all levels

8. Demonstrate control

- Use an appropriate level of formality for tracking and reporting
- Make plans and progress visible to all
- Measure progress through focus on delivery of products rather than completed activities
- Manage proactively

- Evaluate continuing project viability based on the business objectives

Prerequisites for using DSDM

In order for DSDM to be a success, there are 9 instrumental factors that need to be met. If these cannot be met, it presents a risk to the Atern approach which is not necessarily a showstopper but which does need to be managed. The Project Approach Questionnaire also highlights these risks.

1. Acceptance of the Atern philosophy before starting work.
2. Appropriate empowerment of the Solution Development Team.
3. Commitment of senior business management to provide the necessary Business Ambassador (and Business Advisor) involvement.
4. Incremental delivery
5. Access by the Solution Development Team to business roles
6. Solution Development Team stability.
7. Solution Development Team skills.
8. Solution Development Team size.
9. A supportive commercial relationship.

Overview of DSDM version 4.2

As an extension of rapid application development, DSDM focuses on information systems projects that are characterized by tight schedules and budgets. DSDM addresses the most common failures of information systems projects, including exceeding budgets, missing deadlines, and lack of user involvement and top-management commitment. By encouraging the use of RAD, however, careless adoption of DSDM may increase the risk of cutting too many corners. DSDM consists of three phases: pre-project phase, project life-cycle phase, and post-project phase.

A project life-cycle phase is subdivided into 5 stages: feasibility study, business study, functional model iteration, design and build iteration, and implementation.

In some circumstances, there are possibilities to integrate practices from other methodologies, such as Rational Unified Process (RUP), Extreme Programming (XP), and PRINCE2, as complements to DSDM. Another agile method that has some similarity in process and concept to DSDM is Scrum.

In July 2006, DSDM Public Version 4.2 [118] was made available for individuals to view and use; however, anyone reselling DSDM must still be a member of the not-for-profit consortium.

DSDM V4.2 Project Life-cycle
Overview: three phases of DSDM V4.2

The DSDM framework consists of three sequential phases, namely the pre-project, project life-cycle and post-project phases. The project phase of DSDM is the most elaborate of the three phases. The project life-cycle phase consists of 5 stages that form an iterative step-by-step approach in developing an IS. The three phases and corresponding stages are explained extensively in the subsequent sections. For each stage/phase, the most important activities are addressed and the deliverables are mentioned.

Phase 1 - The Pre-project
In the pre-project phase candidate projects are identified, project funding is realized and project commitment is ensured. Handling these issues at an early stage avoids problems at later stages of the project like cows.

Phase 2 - The Project life-cycle
The process overview in the figure below shows the project life-cycle of this phase of DSDM. It depicts the 5 stages a project will have to go through to create an implemented system. The first two stages, the Feasibility Study and Business Study are sequential

phases that complement to each other. After these phases have been concluded, the system is developed iteratively and incrementally in the Functional Model Iteration, Design & Build Iteration and Implementation stages. The iterative and incremental nature of DSDM will be addressed further in a later section.

Phase 3 - Post-project

The post-project phase ensures the system operates effectively and efficiently. This is realized by maintenance, enhancements and fixes according to DSDM principles. The maintenance can be viewed as continuing development based on the iterative and incremental nature of DSDM. Instead of finishing the project in one cycle, usually the project can return to the previous phases or stages so that the previous step and the deliverable products can be refined.

Below is the process-data diagram of DSDM as a whole Project life-cycle with all of its four steps. This diagram depicts the DSDM iterative development, started on functional model iteration, design and build iteration, and implementation phase.

Figure 13-1 shows the he process-data diagram of DSDM Project Life-cycle

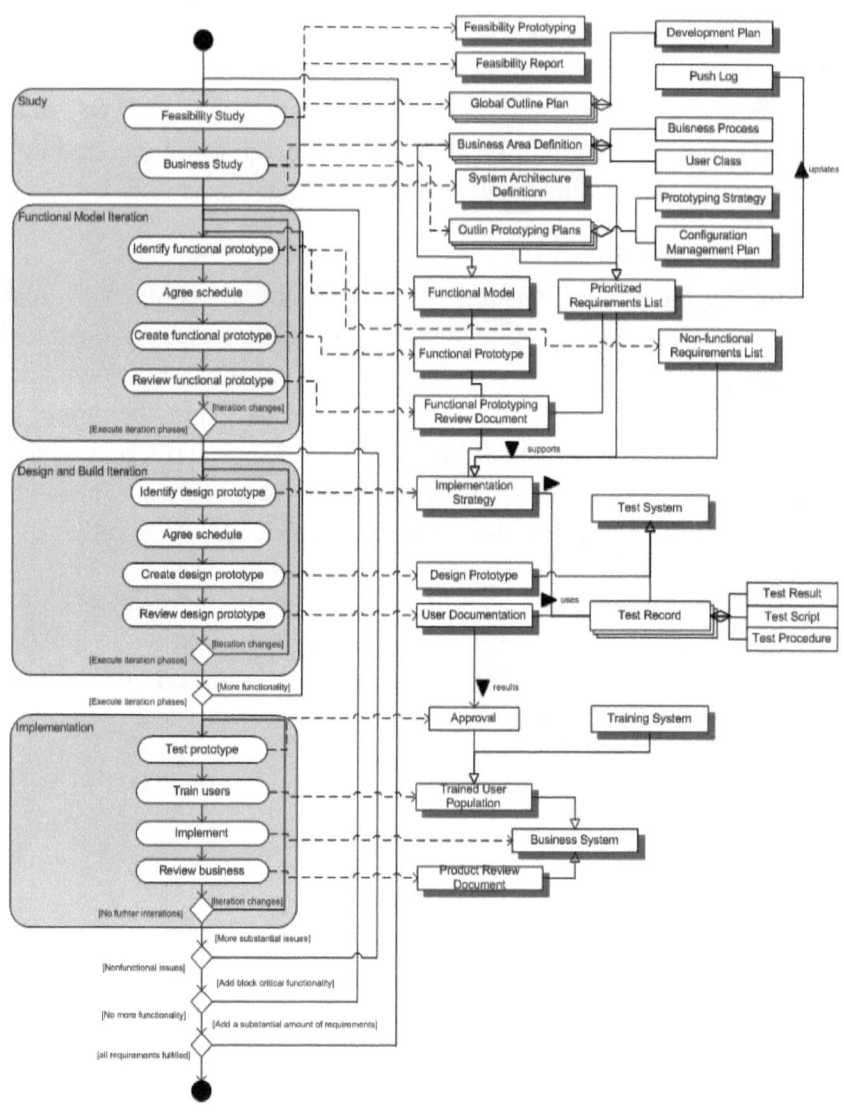

Figure 13-2. The process-data diagram of DSDM Project Life-cycle

Four stages of the DSDM V4.2 Project life-cycle

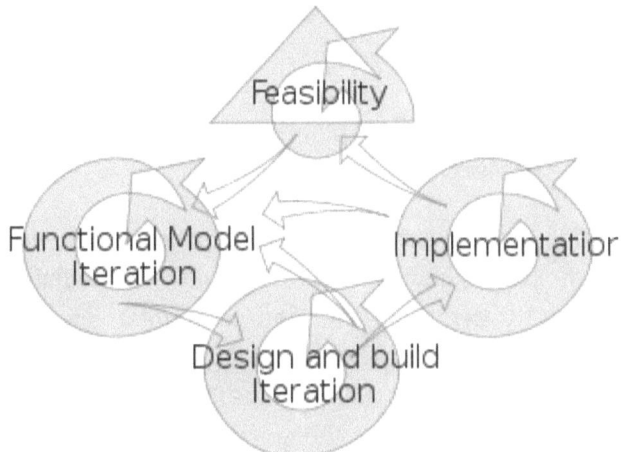

Figure 13-3. Model of the DSDM V4.2 software development process.

Stage 1A: The Feasibility Study
During this stage of the project, the feasibility of the project for the use of DSDM is examined. Prerequisites for the use of DSDM are addressed by answering questions like: 'Can this project meet the required business needs?', 'Is this project suited for the use of DSDM?' and 'What are the most important risks involved?'. The most important techniques used in this phase are the Workshops.

The deliverables for this stage are the Feasibility Report and the Feasibility Prototype that address the feasibility of the project at hand. It is extended with a Global Outline Plan for the rest of the project and a Risk Log that identifies the most important risks for the project.

Stage 1B: The Business Study
The business study extends the feasibility study. After the project has been deemed feasible for the use of DSDM, this stage examines the influenced business processes, user groups involved and their respective needs and wishes. Again the workshops are one of the most valuable techniques, workshops in which the different

stakeholders come together to discuss the proposed system. The information from these sessions is combined into a requirements list. An important property of the requirements list is the fact that the requirements are (can be) prioritized. These requirements are prioritized using the MoSCoW approach. Based on this prioritization, a development plan is constructed as a guideline for the rest of the **project.**

An important project technique used in the development of this plan is timeboxing. This technique is essential in realizing the goals of DSDM, namely being on time and on budget, guaranteeing the desired quality. System architecture is another aid to guide the development of the IS. The deliverables for this stage are a business area definition that describes the context of the project within the company, a system architecture definition that provides an initial global architecture of the IS under development together with a development plan that outlines the most important steps in the development process. At the base of these last two documents, there is the prioritized requirements list. This list states all the requirements for the system, organized according to the MoSCoW principle. Last, the Risk Log is updated with the facts that have been identified during this phase of DSDM.

Stage 2: Functional Model Iteration
The requirements that have been identified in the previous stages are converted to a functional model. This model consists of both a functioning prototype and models. Prototyping is one of the key project techniques within this stage that helps to realize good user involvement throughout the project. Different user groups review the developed prototype. In order to assure quality, testing is implemented throughout every iteration of DSDM. An important part of testing is realized in the Functional Model Iteration. The Functional Model can be subdivided into four sub-stages:

<u>Identify Functional Prototype</u>: Determine the functionalities to be implemented in the prototype that results from this iteration.

Agree Schedule: Agree on how and when to develop these functionalities.

Create Functional Prototype: Develop the prototype. Investigate, refine, and consolidate it with the combined Functional prototype of previous iterations.

Review Prototype: Check the correctness of the developed prototype. This can be done via testing by end-user, then use the test records and user's feedbacks to generate the functional prototyping review document.

The deliverables for this stage are a Functional Model and a Functional Prototype that together represent the functionalities that could be realized in this iteration, ready for testing by users. Next to this, the Requirements List is updated, deleting the items that have been realized and rethinking the prioritization of the remaining requirements. The Risk Log is also updated by having risk analysis of further development after reviewing the prototyping document.

Stage 3: Design and Build Iteration
The focus of this DSDM iteration is to integrate the functional components from the previous phase into one system that satisfies user needs. It also addresses the non-functional requirements that have been set for the IS. Again testing is an important ongoing activity in this stage. The Design and Build Iteration can be subdivided into four sub-stages:

Identify Design Prototype: Identify functional and non-functional requirements that need to be in the tested system.

Agree Schedule: Agree on how and when to realize these requirements.

Create Design Prototype: Create a system that can safely be handed to end-users for daily use. They investigate, refine, and consolidate the prototype of current iteration within prototyping process are also important in this sub-stage.

Review Design Prototype: Check the correctness of the designed system. Again testing and reviewing are the main techniques used, since the test records and user's feedbacks are important to generate the user documentation.

The deliverables for this stage are a Design Prototype during the phase that end users get to test and at the end of the Design and Build Iteration the Tested System is handed over to the next phase. In this stage, the system is mainly built where the design and functions are consolidated and integrated in a prototype. Another deliverable for this stage is a User Documentation.

Stage 4: Implementation

In the Implementation stage, the tested system including user documentation is delivered to the users and training of future users is realized. The system to be delivered has been reviewed to include the requirements that have been set in the beginning stages of the project. The Implementation stage can be subdivided into four sub-stages:

User Approval and Guidelines: End users approve the tested system for implementation and guidelines with respect to the implementation and use of the system are created.

Train Users: Train future end user in the use of the system.

Implement: Implement the tested system at the location of the end users.

Review Business: Review the impact of the implemented system on the business, a central issue will be whether the system meets the goals set at the beginning of the project. Depending on this the project goes to the next phase, the post-project or loops back to one

Table 13-1. The four steps of the DSDM V4.2 Project Life-cycle

Activity	Sub activity	Description
Study	Feasibility Study	Stage where the suitability of DSDM is assessed. Judging by the type of project, organizational and people issues, the decision is made, whether to use DSDM or not. Therefore, it will generate a FEASIBILITY REPORT, a FEASIBILITY PROTOTYPE, and a GLOBAL OUTLINE PLAN, which includes a DEVELOPMENT PLAN and a RISK LOG.
	Business Study	Stage where the essential characteristics of business and technology are analyzed. Approach to organize workshops, where a sufficient number of the customer's experts are gathered to be able to consider all relevant facets of the system, and to be able to agree on development priorities. In this stage, a PRIORITIZED REQUIREMENTS LIST, a BUSINESS AREA DEFINITION, a SYSTEM ARCHITECTURE DEFINITION, and an OUTLINE PROTOTYPING PLAN are developed.
Functional Model Iteration	Identify functional prototype	Determine the functionalities to be implemented in the prototype that results from this iteration. In this sub-stage, a FUNCTIONAL MODEL is developed according to the deliverables result of business study stage.
	Agree schedule	Agree on how and when to develop these functionalities.

Activity	Sub activity	Description
	Create functional prototype	Develop the FUNCTIONAL PROTOTYPE, according to the agreed schedule and FUNCTIONAL MODEL.
	Review functional prototype	Check correctness of the developed prototype. This can be done via testing by end-user and/or reviewing documentation. The deliverable is a FUNCTIONAL PROTOTYPING REVIEW DOCUMENT.
Design and Build Iteration	Identify design prototype	Identify functional and non-functional requirements that need to be in the tested system. And based on these identifications, an IMPLEMENTATION STRATEGY is involved. If there is a TEST RECORD from the previous iteration, then it will be also used to determine the IMPLEMENTATION STRATEGY.
	Agree schedule	Agree on how and when to realize these requirements.
	Create design prototype	Create a system (DESIGN PROTOTYPE) that can safely be handed to end-users for daily use, also for testing purposes.
	Review design prototype	Check the correctness of the designed system. Again, testing and reviewing are the main techniques used. An USER DOCUMENTATION and a TEST RECORD will be developed.
Implementation	User approval and guidelines	End users approve the tested system (APPROVAL) for implementation and guidelines with respect to the implementation and use of the system

Activity	Sub activity	Description
		are created.
	Train users	Train future end user in the use of the system. TRAINED USER POPULATION is the deliverable of this sub-stage.
	Implement	Implement the tested system at the location of the end users, called as DELIVERED SYSTEM.
	Review business	Review the impact of the implemented system on the business, a central issue will be whether the system meets the goals set at the beginning of the project. Depending on this, the project goes to the next stage, the post-project or loops back to one of the preceding stages for further development. This review is will be documented in a PROJECT REVIEW DOCUMENT.

DSDM V4.2 Functional Model Iteration
Meta-data model

The associations between concepts of deliverables in Functional Model Iteration stage are depicted in the meta-data model below. This meta-data model will be combined with the meta-process diagram of Functional Model Iteration phase in the next part.

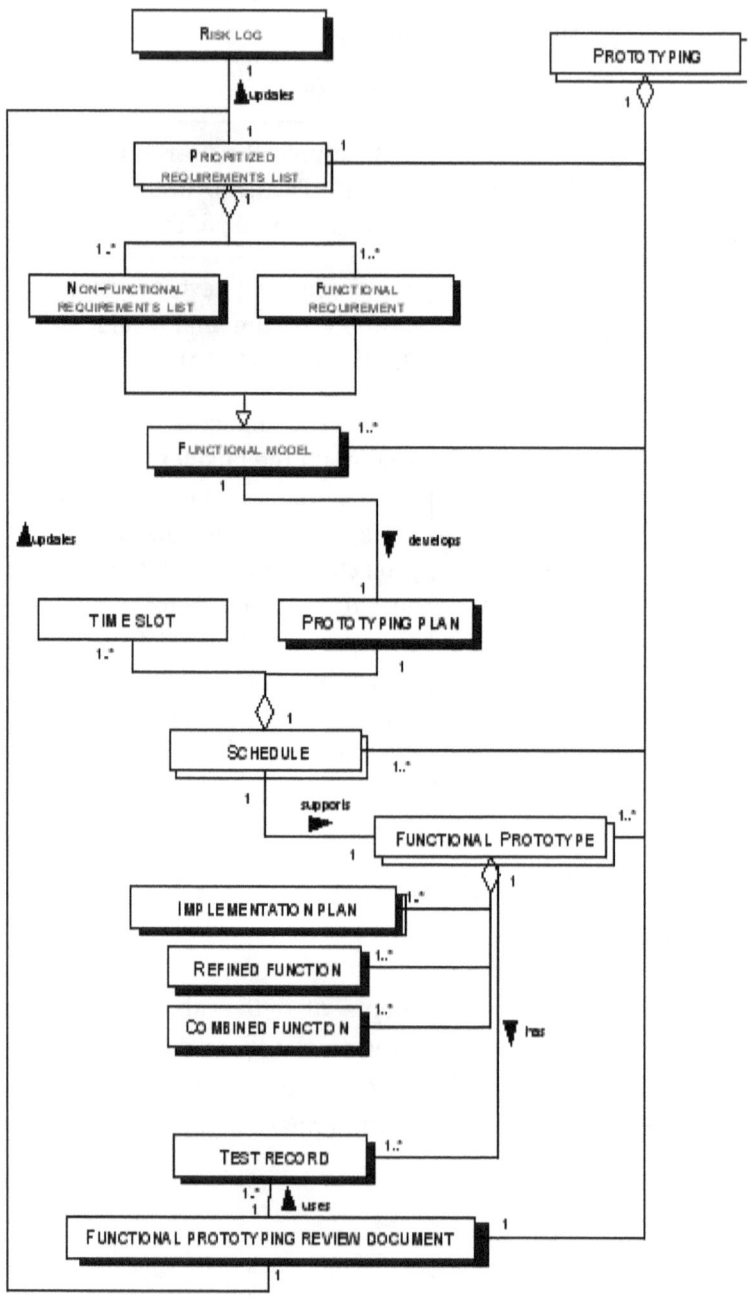

Figure 13-4. Meta-data model of Functional Model Iteration

Concept	Definition
RISK LOG	Log of identified risk. Important since the next stage onward, encountered problem will be more difficult to address. This risk log will need to be updated continuously. (VTT Publication 478)
PRIORITIZED REQUIREMENTS LIST	List of requirements based on its prioritization. The prioritization process is based on MoSCoW technique, to determine which requirements must be implemented first into the system (the ones that meet the business needs), and so on.
NON-FUNCTIONAL REQUIREMENTS LIST	List of non-functional requirements is mainly to be dealt in the next stage. (VTT Publication 478)
FUNCTIONAL REQUIREMENT	Function that is used to build the model and prototyping according to its priority.
FUNCTIONAL MODEL	Model that is built according to the functional requirements. It will be used in order to develop the functional prototype in the next sub-stage. This concept will be used to develop a PROTOTYPING PLAN.
PROTOTYPING	The process of quickly putting together a working model (a prototype) in order to test various aspects of the design, illustrate ideas or features and gather early user feedback.
TIME SLOT	The list of available times to do certain activities in order to perform the plan according to the schedule.
PROTOTYPING PLAN	Activities plan within prototyping process that will be performed in available time slots according to the schedule.

Concept	Definition
SCHEDULE	A set of activities plan with the related time agreed by the developers. This concept will be used to support the implementation of FUNCTIONAL PROTOTYPE.
FUNCTIONAL PROTOTYPE	A prototype of the functions the system should perform and how it should perform them.
IMPLEMENTATION PLAN	A preparation of activities to implement the functional prototyping according to the schedule and the prioritized requirements list.
REFINED FUNCTION	Function of prototype that is being refined within current iteration before it is combined to the others and tested.
COMBINED FUNCTION	Function of prototyped that is combined with the other functional prototypes of previous iteration. The new combination functional prototype will be tested in the next stage.
TEST RECORD	Record set of testing where the test script, test procedure, and test result are included. This test record is used to develop the FUNCTIONAL PROTOTYPING REVIEW DOCUMENT, and is also used indirectly to update the PRIORITIZED REQUIREMENTS LIST.
FUNCTIONAL PROTOTYPING REVIEW DOCUMENT	It collects the users' comments about the current increment, working as input for subsequent iterations (VTT Publication 478). This review document will be used to update the RISK LOG and PRIORITIZED REQUIREMENTS LIST.

Process-data model

Identify functional prototype activity is to identify the functionalities that would be in the prototype of current iteration. Recall that both, analysis and coding are done; prototypes are built, and the experiences gained from them are used in improving the analysis models (based also on updated prioritized requirements list and updated risk log). The built prototypes are not to be entirely discarded, but gradually steered towards such quality that they can be included in the final system. Agree schedule is to determine when and how the prototyping will be implemented; it extends the scope to the available timetable and prototyping plan. Moreover, since testing is implemented throughout the whole process, it is also an essential part of this phase, and therefore it is included in the Review Prototype activity right after the functional prototype is built and the test record will eventually be used in the review prototype process and generates the review document. Below is the process-data diagram of Functional Model Iteration stage.

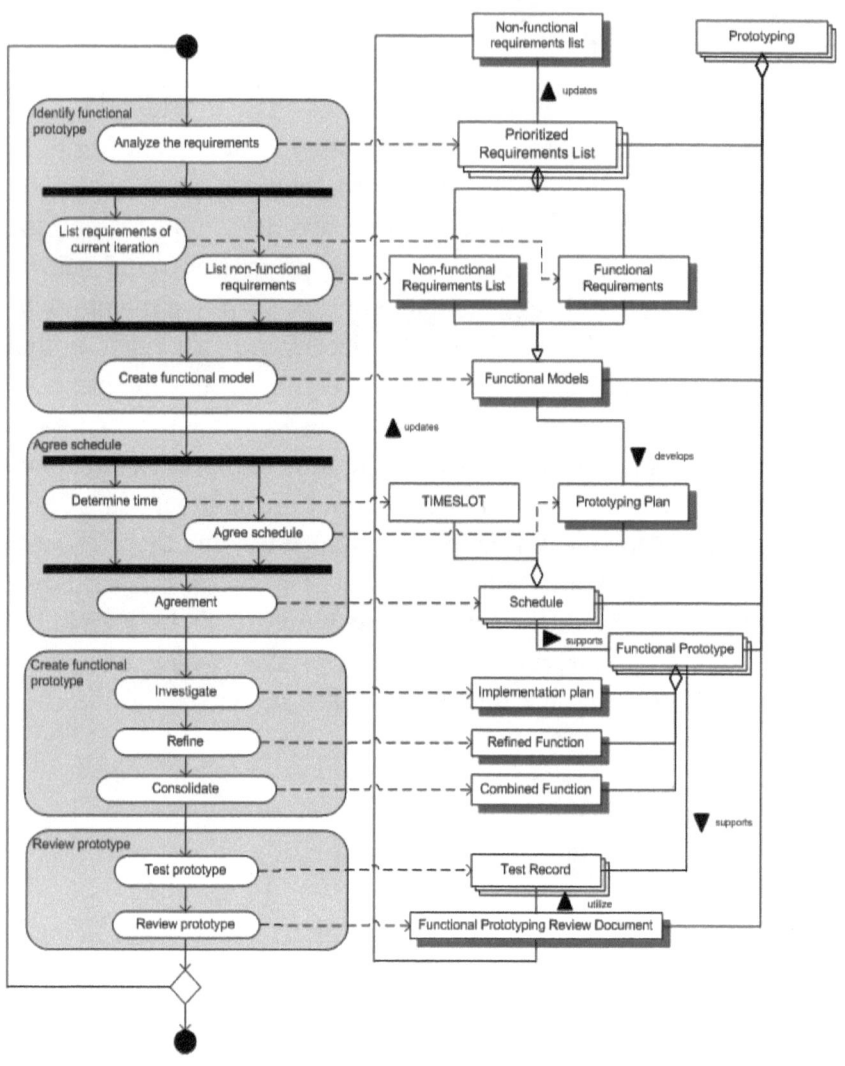

Figure 13-5. Model of the Functional Model Iteration.

Activity	Sub activity	Description
Identify functional prototype	Analyze the requirements	The requirements of current prototype are analyzed according to the prioritized requirements list that is previously developed (in previous iteration and/or in previous phase, which is business study phase).

Activity	Sub activity	Description
	List requirements of current iteration	Select the functional requirements that would be implemented in the current iteration's prototype, and list them in the FUNCTIONAL REQUIREMENT.
	List non-functional requirements	List the non-functional requirements of the system that is being developed in NON-FUNCTIONAL REQUIREMENTS LIST.
	Create functional model	Analysis model and prototype codes are included in this sub-activity to develop the FUNCTIONAL MODEL.
Agree schedule	Determine time	Identify possible timetable (TIME SLOT) to perform the prototyping activities according to the prototyping plan and prototyping strategy.
	Design development	The PROTOTYPING PLAN, including all prototyping activities that will be performed on available TIME SLOT.
	Agreement	The agreement SCHEDULE of when and how the prototyping activities should be performed.
Create functional prototype	Investigate	Investigate the requirements; analyze the FUNCTIONAL MODEL that has been built in earlier activity, and set the IMPLEMENTATION PLAN according to the analysis model, and will be used to build the prototype in the next sub-activity.
	Refine	Implement the FUNCTIONAL MODEL and IMPLEMENTATION PLAN to build a FUNCTIONAL PROTOTYPE. This prototype will be then refined before it is combined to the other functions. This prototype will

Activity	Sub activity	Description
		be gradually steered towards such quality that it can be included in the final system (through refining process).
	Consolidate	Consolidate the refined FUNCTIONAL PROTOTYPE with the other prototype of previous iteration. The new combination FUNCTIONAL PROTOTYPE will be tested in the next activity.
Review prototype	Test prototype	The essential part of DSDM that needs to be included throughout the whole process of DSDM. The TEST RECORD will be used together with users' comments to develop the PROTOTYPING REVIEW DOCUMENT in the next activity of FMI phase.
	Review prototype	Collects the user comments and documentation. The test records will play an important role to develop this review report. Based on this FUNCTIONAL PROTOTYPING REVIEW DOCUMENT, the prioritized requirements list and risk log will be updated, and decide to set the next course whether or not to do another iteration of FMI phase.

Notes

---

[1] Agile software development is a group of software development methodologies based on iterative and incremental development, where requirements and solutions evolve through collaboration between self-organizing, cross-functional teams. It promotes adaptive planning, evolutionary development and delivery, a time-boxed iterative approach, and encourages rapid and flexible response to change. It is a conceptual

framework that promotes foreseen interactions throughout the development cycle.

² MoSCoW is a prioritization technique used in business analysis and software development to reach a common understanding with stakeholders on the importance they place on the delivery of each requirement - also known as MoSCoW prioritization or MoSCoW analysis. According to A Guide to the Business Analysis Body of Knowledge, version 2.0, section 6.5.1.2, the MoSCoW categories are as follows: M - MUST: Describes a requirement that must be satisfied in the final solution for the solution to be considered a success. S - SHOULD: Represents a high-priority item that should be included in the solution if it is possible. This is often a critical requirement but one which can be satisfied in other ways if strictly necessary. C - COULD: Describes a requirement that is considered desirable but not necessary. This will be included if time and resources permit. W - WON'T: Represents a requirement that stakeholders have agreed will not be implemented in a given release, but may be considered for the future. The o's in MoSCoW are added simply to make the word pronounceable, and are often left lower case to indicate that they do not stand for anything. MOSCOW is an acceptable variant, with the 'o's in upper case.

³ The Arctic Tern (*Sterna paradisaea*) is a seabird of the tern family Sternidae. This bird has a circumpolar breeding distribution covering the Arctic and sub-Arctic regions of Europe, Asia, and North America (as far south as Brittany and Massachusetts). The species is strongly migratory, seeing two summers each year as it migrates from its northern breeding grounds along a winding route to the oceans around Antarctica and back, a round trip of about 70,900 km (c. 44,300 miles) each year. This is by far the longest regular migration by any known animal. The Arctic Tern flies as well as glides through the air, performing almost all of its tasks in the air. It nests once every one to three years (depending on its mating cycle); once it has finished nesting, it takes to the sky for another long southern migration.

⁴ Extreme Programming (XP) is a software development methodology, which is intended to improve software quality and responsiveness to changing customer requirements. As a type of agile software development, it advocates frequent "releases" in short development cycles (timeboxing), which is intended to improve productivity and introduce checkpoints where new customer requirements can be adopted.

[5] Timeboxing is a planning technique common in planning projects (typically for software development), where the schedule is divided into a number of separate periods (timeboxes, normally two to six weeks long), with each part having its own deliverables, deadline and budget. Timeboxing is a core aspect of rapid application development (RAD) software development processes such as dynamic systems development method (DSDM) and agile software development.

[6] Test-driven development (TDD) is a software development process that relies on the repetition of a very short development cycle: first the developer writes a failing automated test case that defines a desired improvement or new function, then produces code to pass that test and finally refactors the new code to acceptable standards. Kent Beck, who is credited with having developed or 'rediscovered' the technique, stated in 2003 that TDD encourages simple designs and inspires confidence.

Chapter 14. Place/ Transition Net

Also known as Petri Nets, this conceptual modeling technique allows a system to be constructed with elements that can be described by direct mathematical means. The Petri net, because of its nondeterministic execution properties and well defined mathematical theory, is a useful technique for modeling concurrent system behavior, i.e. simultaneous process executions.

A **Petri net** (also known as a **place/transition net** or **P/T net**) is one of several mathematical modeling languages for the description of distributed systems. A Petri net is a directed bipartite graph, in which the nodes represent transitions (i.e. events that may occur, signified by bars) and places (i.e. conditions, signified by circles). The directed arcs describe which places are pre- and/or post conditions for which transitions (signified by arrows). Some sources [119] state that Petri nets were invented in August 1939 by Carl Adam Petri—at the age of 13—for describing chemical processes.

Like industry standards such as UML activity diagrams, BPMN[1] and EPCs, Petri nets offer a graphical notation for stepwise processes that include choice, iteration, and concurrent execution. Unlike these standards, Petri nets have an exact mathematical definition of their execution semantics, with a well-developed mathematical theory for process analysis.

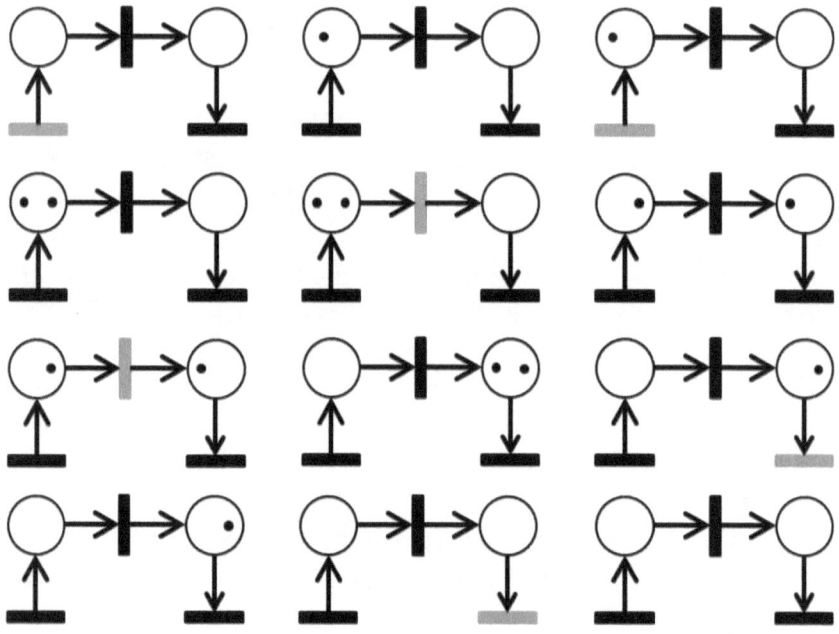

Figure 14-1. Successive Petri net trajectory example

Petri net basics

A Petri net consists of *places*, *transitions*, and *directed arcs*. Arcs run from a place to a transition or vice versa, never between places or between transitions. The places from which an arc runs to a transition are called the *input places* of the transition; the places to which arcs run from a transition are called the *output places* of the transition.

Places may contain a natural number of tokens. A distribution of tokens over the places of a net is called a *marking*. A transition of a Petri net may *fire* whenever there is a token at the start of all input arcs; when it fires, it consumes these tokens, and places tokens at the end of all output arcs. A firing is atomic, i.e., a single non-interruptible step.

Execution of Petri nets is nondeterministic: when multiple transitions are enabled at the same time, any one of them may fire. If a transition is enabled, it may fire, but it does not have to.

Since firing is nondeterministic, and multiple tokens may be present anywhere in the net (even in the same place), Petri nets are well suited for modeling the concurrent behavior of distributed systems.

Formal definition and basic terminology

The following formal definition is loosely based on Peterson, 1981 [120]. Many alternative definitions exist.

Syntax

A **Petri net graph** (called *Petri net* by some, but see below) is a 3-tuple (S, T, W), where

- S is a finite set of places
- T is a finite set of transitions
- S and T are disjoint, i.e. no object can be both a place and a transition
- $W: (S \times T) \cup (T \times S) \to \mathbb{N}$ is a multiset of arcs[2], i.e. it defines arcs and assigns to each arc a non-negative integer *arc multiplicity*; note that no arc may connect two places or two transitions.

The *flow relation* is the set of arcs: $F = \{(x, y) | W(x, y) > 0\}$. In many textbooks, arcs can only have multiplicity 1. These texts often define Petri nets using F instead of W. When using this convention, a Petri net graph is a bipartite multigraph $(S \cup T, F)$ with node partitions S and T.

The *preset* of a transition t is the set of its *input places*: $^\bullet t = \{s \in S | W(s, t) > 0\}$; its *postset* is the set of its *output places*: $t^\bullet = \{s \in S | W(t, s) > 0\}$. Definitions of pre- and postsets of places are analogous.

A *marking* of a Petri net (graph) is a multiset of its places, i.e., a mapping $M: S \to \mathbb{N}$. We say the marking assigns to each place a number of *tokens*.

A **Petri net** (called *marked Petri net* by some, see above) is a 4-tuple (S, T, W, M_0), where

- (S, T, W) is a Petri net graph;
- M_0 is the *initial marking*, a marking of the Petri net graph.

Execution semantics

The behavior of a Petri net is defined as a relation on its markings, as follows.

Note that markings can be added like any multiset: $M + M' \triangleq \{s \to M(s) + M'(s) | s \in S\}$

The execution of a Petri net graph $G = (S, T, W)$ can be defined as the *transition relation* $\to G$ on its markings, as follows:

- for any t in T:
$$M \to_{G,t} M' \stackrel{D}{\Leftrightarrow} \exists M: S \to \mathbb{N}: M = M + \sum_{s \in S} W(s,t) \land M'$$
$$= M'' + \sum_{s \in S} W(t,s)$$
- $M \to_G M' \stackrel{D}{\Leftrightarrow} \exists t \in T: M \to_{G,t} M'$

In words:

- firing a transition t in a marking M consumes $W(s,t)$ tokens from each of its input places s, and produces $W(t,s)$ tokens in each of its output places s
- a transition is *enabled* (it may *fire*) in M if there are enough tokens in its input places for the consumptions to be possible, i.e. iff $\forall s: M(s) \geq W(s,t)$.

We are generally interested in what may happen when transitions may continually fire in arbitrary order.

We say that a marking M' *is reachable from* a marking M *in one step* if $M \rightarrow_G M'$; we say that it *is reachable from* M if $M \rightarrow_G^* M'$, where $\rightarrow_G^*$ is the reflexive transitive closure of $\rightarrow_G$; that is, if it is reachable in 0 or more steps.

For a (marked) Petri net $N = (S, T, W, M_0)$, we are interested in the firings that can be performed starting with the initial marking M_0. Its set of *reachable markings* is the set $R(N) \triangleq \{M' | M_0 \rightarrow_{(S,T,W)}^* M'\}$

The *reachability graph* of N is the transition relation $\rightarrow_G$ restricted to its reachable markings $R(N)$. It is the state space[3] of the net.

A *firing sequence* for a Petri net with graph G and initial marking M_0 is a sequence of transitions $\vec{\sigma} = \langle t_{i1} \ldots t_{in} \rangle$ such that $M_0 \rightarrow_{G, t_{i1}} M_1 \wedge \ldots \wedge M_{n-1} \rightarrow_{G, t_{in}} M_n$. The set of firing sequences is denoted as $L(N)$.

Variations on the definition

As already remarked, a common variation is to disallow arc multiplicities and replace the bag of arcs W with a simple set, called the *flow relation*, $F \subseteq (S \times T) \cup (T \times S)$. This does not limit expressive power as both can represent each other.

Another common variation, e.g. in, e.g. Desel and Juhás (2001) [121], is to allow *capacities* to be defined on places. This is discussed under *extensions* below.

Formulation in terms of vectors and matrices

The markings of a Petri net (S, T, W, M_0) can be regarded as vectors of nonnegative integers of length $|S|$.

Its transition relation can be described as a pair of $|S|$ by $|T|$ matrices:

- W^-, defined by $\forall s, t: W^-[s, t] = W(s, t)$

- W^+, defined by $\forall s, t: W^+[s,t] = W(t,s)$

Then their difference

- $W^T = W^+ - W^-$

can be used to describe the reachable markings in terms of matrix multiplication, as follows. For any sequence of transitions w, write $o(w)$ for the vector that maps every transition to its number of occurrences in w. Then, we have

$R(N) = \{M | \exists w: M = M_0 + W^T \cdot o(w) \wedge w \text{ is a firing sequence of } N\}$.

Note that it must be required that w is a firing sequence; allowing arbitrary sequences of transitions will generally produce a larger set.

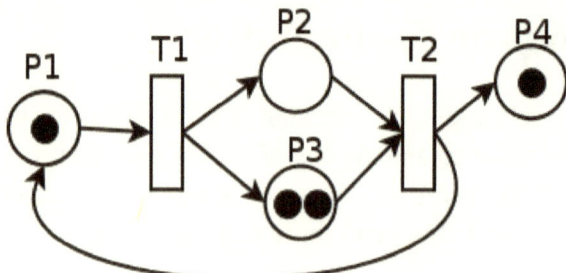

Figure 14-2. Petri net Example

$$W^+ = \begin{bmatrix} * & t1 & t2 \\ p1 & 0 & 1 \\ p2 & 1 & 0 \\ p3 & 1 & 0 \\ p4 & 0 & 1 \end{bmatrix}$$

$$W^- = \begin{bmatrix} * & t1 & t2 \\ p1 & 1 & 0 \\ p2 & 0 & 1 \\ p3 & 0 & 1 \\ p4 & 0 & 0 \end{bmatrix}$$

$$W^t = \begin{bmatrix} * & t1 & t2 \\ p1 & -1 & 1 \\ p2 & 1 & -1 \\ p3 & 1 & -1 \\ p4 & 0 & 1 \end{bmatrix}$$

$$M_0 = \begin{bmatrix} 1 & 0 & 2 & 1 \end{bmatrix}$$

Mathematical properties of Petri nets

One thing that makes Petri nets interesting is that they provide a balance between modeling power and analyzability: many things one would like to know about concurrent systems can be automatically determined for Petri nets, although some of those things are very expensive to determine in the general case. Several subclasses of Petri nets have been studied that can still model interesting classes of concurrent systems, while these problems become easier.

An overview of such decision problems[4], with decidability and complexity results for Petri nets and some subclasses, can be found in Esparza and Nielsen (1995) [122].

Reachability

The reachability[5] problem for Petri nets is to decide, given a Petri net N and a marking M, whether $M \in R(N)$.

Clearly, this is a matter of walking the reachability graph defined above, until we reach the requested marking or we know it can no longer be found. This is harder than it may seem at first: the reachability graph is generally infinite, and it is not easy to determine when it is safe to stop.

In fact, this problem was shown to be EXPSPACE[6]-hard [123] years before it was shown to be decidable at all (Mayr, 1981). Papers continue to be published on how to do it efficiently [124].

While reachability seems to a be a good tool to find erroneous states, for practical problems the constructed graph usually has far too many states to calculate. To alleviate this problem, linear temporal logic[7] is usually used in conjunction with the tableau method[8] to prove that such states cannot be reached. LTL uses the semi-decision technique to find if indeed a state can be reached, by finding a set of necessary conditions for the state to be reached then proving that those conditions cannot be satisfied.

Liveness

Petri nets can be described as having different degrees of liveness $L_1 - L_4$. A Petri net (N, M_0) is called L_k-live iff all of its transitions are L_k-live, where a transition is

- *dead*, iff it can never fire, i.e. it is not in any firing sequence in $L(N, M_0)$
- L_1-live (*potentially fireable*), iff it may fire, i.e. it is in some firing sequence in $L(N, M_0)$
- L_2-live iff it can fire arbitrarily often, i.e. if for every positive integer k, it occurs at least k times in some firing sequence in $L(N, M_0)$
- L_3-live iff it can fire infinitely often, i.e. if for every positive integer k, it occurs at least k times in V, for some prefix-closed set of firing sequences $V \subseteq L(N, M_0)$
- L_4-live (*live*) iff it may always fire, i.e., it is L_1-live in every reachable marking in $R(N, M_0)$

Note that these are increasingly stringent requirements: L_{j+1}-liveness implies L_j-liveness, for $j \in 1,2,3$.

These definitions are in accordance with Murata's overview [125], which additionally uses L_0-*live* as a term for *dead*.

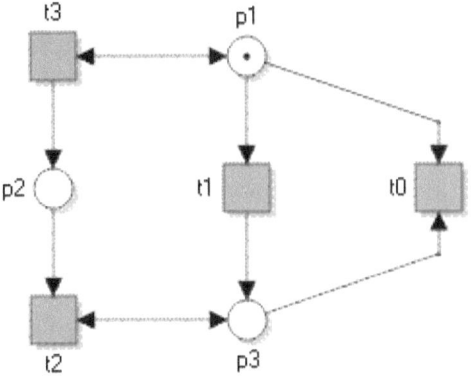

Figure 14-3. A Petri net in which transition t_0 is dead, and $\forall j > 0: t_j$ is L_j – live

Boundedness

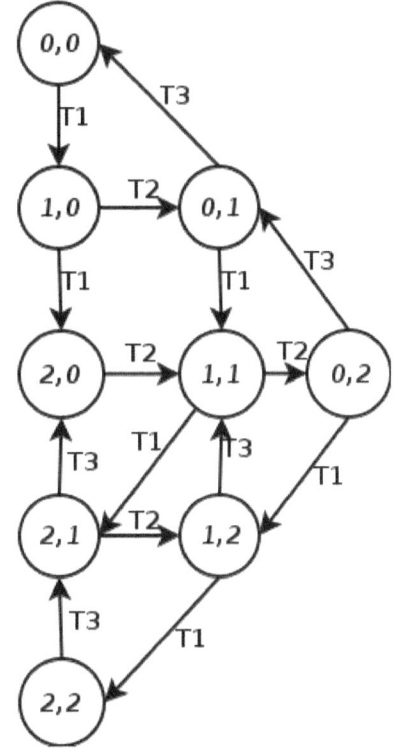

Figure 14-4. The reachability graph of N2.

A place in Petri net is called *k-bounded* if it does not contain more than *k* tokens in all reachable markings, including the initial marking; it is *safe* if it is 1-bounded; it is *bounded* if it is *k-bounded* for some *k*.

A (marked) Petri net is called *k-bounded*, *safe*, or *bounded* when all of its places are. A Petri net (graph) is called *(structurally) bounded* if it is bounded for every possible initial marking.

Note that a Petri net is bounded if and only if its reachability graph is finite.

Boundedness is decidable by looking at covering[9], by constructing the Karp–Miller Tree [126].

It can be useful to impose explicitly a bound on places in a given net. This can be used to model limited system resources.

Some definitions of Petri nets explicitly allow this as a syntactic feature [127]. Formally, *Petri nets with place capacities* can be defined as tuples (S, T, W, C, M_0), where (S, T, W, M_0) is a Petri net, $C: P \rightarrow IN$ an assignment of capacities to (some or all) places, and the transition relation is the usual one restricted to the markings in which each place with a capacity has at most that many tokens.

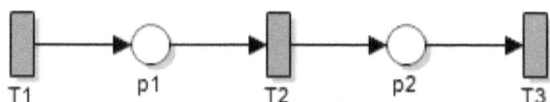

Figure 14-5. An unbounded Petri net, *N*.

For example, if in the net *N*, both places are assigned capacity 2, we obtain a Petri net with place capacities, say *N2*; its reachability graph is displayed on the right.

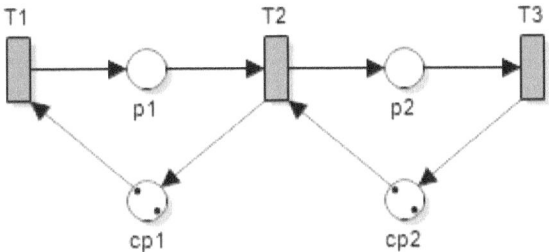

Figure 14-6. A two-bounded Petri net, obtained by extending N with "counter-places".

Alternatively, places can be made bounded by extending the net. To be exact, a place can be made *k*-bounded by adding a "counter-place" with flow opposite to that of the place, and adding tokens to make the total in both places *k*.

Discrete, continuous, and hybrid Petri nets

As well as discrete events, there are Petri nets for continuous and hybrid discrete-continuous processes and useful in discrete, continuous and hybrid control theory[10] [128] and related to discrete, continuous and hybrid automata[11].

Extensions

There are many extensions to Petri nets. Some of them are completely backwards-compatible (e.g. colored Petri nets) with the original Petri net, some add properties that cannot be modeled in the original Petri net (e.g. timed Petri nets). If they can be modeled in the original Petri net, they are not real extensions, instead, they are convenient ways of showing the same thing, and can be transformed with mathematical formulas back to the original Petri net, without losing any meaning. Extensions that cannot be transformed are sometimes very powerful, but usually lack the full range of mathematical tools available to analyze normal Petri nets.

The term high-level Petri net is used for many Petri net formalisms that extend the basic P/T net formalism; this includes colored Petri nets, hierarchical Petri nets, and all other extensions sketched in this section. The term is also used specifically for the type of colored nets supported by CPN Tools.

A short list of possible extensions:

- Additional types of arcs; two common types are:
  o a *reset arc* does not impose a precondition on firing, and empties the place when the transition fires; this makes reachability undecidable [129], while some other properties, such as termination, remain decidable [130];
  o an *inhibitor arc* imposes the precondition that the transition may only fire when the place is empty; this allows arbitrary computations on numbers of tokens to be expressed, which makes the formalism Turing complete[12].
- In a standard Petri net, tokens are indistinguishable. In a Colored Petri Net, every token has a value [131]. In popular tools for colored Petri nets such as CPN Tools, the values of tokens are typed, and can be tested (using *guard* expressions) and manipulated with a functional programming language. A subsidiary of colored Petri nets are the well-formed Petri nets, where the arc and guard expressions are restricted to make it easier to analyze the net.
- Another popular extension of Petri nets is hierarchy: Hierarchy in the form of different views supporting levels of refinement and abstraction were studied by Fehling. Another form of hierarchy is found in so-called object Petri nets or object systems where a Petri net can contain Petri nets as its tokens inducing a hierarchy of nested Petri nets that communicate by synchronization of transitions on different levels.
- A Vector Addition System with States (VASS) can be seen as a generalization of a Petri net. Consider a finite state automaton[13] where each transition is labeled by a transition from the Petri net. The Petri net is then synchronized with the finite state automaton, i.e., a transition in the automaton is taken at the

same time as the corresponding transition in the Petri net. It is only possible to take a transition in the automaton if the corresponding transition in the Petri net is enabled, and it is only possible to fire a transition in the Petri net if there is a transition from the current state in the automaton labeled by it. (The definition of VASS is usually formulated slightly differently.)
- Prioritized Petri nets add priorities to transitions, whereby a transition cannot fire, if a higher-priority transition is enabled (i.e. can fire). Thus, transitions are in priority groups, and e.g. priority group 3 can only fire if all transitions are disabled in groups 1 and 2. Within a priority group, firing is *still* non-deterministic.
- The nondeterministic property has been a very valuable one, as it lets the user abstract a large number of properties (depending on what the net is used for). In certain cases, however, the need arises to model also the timing, not only the structure of a model. For these cases, timed Petri nets have evolved, where there are transitions that are timed, and possibly transitions which are not timed (if there are, transitions that are not timed have a higher priority than timed ones). A subsidiary of timed Petri nets are the stochastic Petri nets that add nondeterministic time through adjustable randomness of the transitions. The exponential random distribution is usually used to 'time' these nets. In this case, the nets' reachability graph can be used as a Markov chain[14].
- Dualistic Petri Nets (dP-Nets) is a Petri Net extension developed by E. Dawis, et al. [132] to represent better real-world process. dP-Nets balance the duality of change/no-change, action/passivity, (transformation) time/space, etc., between the bipartite Petri net constructs of transformation and place resulting in the unique characteristic of *transformation marking*, i.e., when the transformation is "working" it is marked. This allows the transformation to fire (or be marked) multiple times representing the real-world behavior of process throughput. Marking of the transformation assumes that transformation

time must be greater than zero. A zero transformation time used in many typical Petri nets may be mathematically appealing but impractical in representing real-world processes. dP-Nets also exploit the power of Petri nets' hierarchical abstraction to depict Process architecture. Complex process systems are modeled as a series of simpler nets interconnected through various levels of hierarchical abstraction. The process architecture of a packet switch is demonstrated in [133], where development requirements are organized around the structure of the designed system. dP-Nets allow any real-world process, such as computer systems, business processes, traffic flow, etc., to be modeled, studied, and improved.

There are many more extensions to Petri nets; however, it is important to keep in mind, that as the complexity of the net increases in terms of extended properties, the harder it is to use standard tools to evaluate certain properties of the net. For this reason, it is a good idea to use the simplest net type possible for a given modeling task.

Restrictions

Instead of extending the Petri net formalism, we can also look at restricting it, and look at particular types of Petri nets, obtained by restricting the syntax in a particular way. The following types are commonly used and studied:

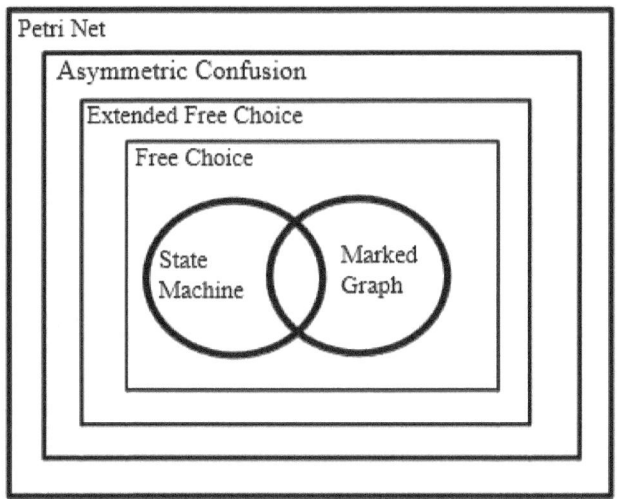

Figure 14-7. Petri net types graphically

1. In a state machine (SM), every transition has one incoming arc, and one outgoing arc. This means, that there cannot be *concurrency*, but there can be *conflict* (i.e. Where should the token from the place go? To one transition or the other?). Mathematically: $\forall t \in T: |t^\bullet| = |^\bullet t| = 1$.
2. In a marked graph (MG), every place has one incoming arc, and one outgoing arc. This means, that there cannot be *conflict*, but there can be *concurrency*. Mathematically: $\forall s \in S: |s^\bullet| = |^\bullet s| = 1$
3. In a *free choice* net (FC), - every arc is either the only arc going from the place, or it is the only arc going to a transition. For instance, there can be *both concurrency and conflict, but not at the same time*. Mathematically: $\forall s \in S: (|s^\bullet| \leq 1) \vee (^\bullet(s^\bullet) = \{s\})$
4. Extended free choice (EFC) – a Petri net that can be *transformed into an FC*.
5. In an asymmetric choice net (AC), concurrency and conflict (in sum, confusion) may occur, but not symmetrically. Mathematically: $\forall s_1, s_2 \in S: (s_1^\bullet \cap s_2^\bullet \neq \emptyset) \to [(s_1^\bullet \subseteq s_2^\bullet) \vee (s_2^\bullet \subseteq s_1^\bullet)]$

Other models of concurrency

Other ways of modeling concurrent computation have been proposed, including process algebra, the actor model, and trace theory [134]. Different models provide tradeoffs of concepts such as compositionality, modularity, and locality.

An approach to relating some of these models of concurrency is proposed in the chapter by Winskel and Nielsen [135].

Application areas

- Software design
- Workflow management
- Process Modeling
- Data analysis
- Concurrent programming
- Reliability engineering
- Diagnosis
- Discrete process control
- Simulation
- KPN modeling

Notes

[1] Business Process Model and Notation (BPMN) is a graphical representation for specifying business processes in a business process model

[2] Sometimes a directed graph or digraph is called a simple digraph to distinguish it from a directed multigraph, in which the arcs constitute a multiset, rather than a set, of ordered pairs of vertices. Also, in a simple digraph loops are disallowed. (A loop is an arc that pairs a vertex to itself.) On the other hand, some texts allow loops, multiple arcs, or both in a digraph.

3 In the theory of discrete dynamical systems, a state space is a directed graph where each possible state of a dynamical system is represented by a vertex, and there is a directed edge from a to b if and only if $f(a) = b$ where the function f defines the dynamical system. State spaces are useful in computer science as a simple model of machines. Formally, a state space can be defined as a tuple [N, A, S, G] where: N is a set of states; A is a set of arcs connecting the states; S is a nonempty subset of N that contains start states; G is a nonempty subset of N that contains the goal states.

4 In computability theory and computational complexity theory, a decision problem is a question in some formal system with a yes-or-no answer, depending on the values of some input parameters. For example, the problem "given two numbers x and y, does x evenly divide y?" is a decision problem. The answer can be either 'yes' or 'no', and depends upon the values of x and y.

5 In graph theory, reachability is the notion of being able to get from one vertex in a directed graph to some other vertex. Note that reachability in undirected graphs is trivial — it is sufficient to find the connected components in the graph, which can be done in linear time.

6 In complexity theory, EXPSPACE is the set of all decision problems solvable by a deterministic Turing machine in $O(2^{p(n)})$ space, where $p(n)$ is a polynomial function of n. (Some authors restrict $p(n)$ to be a linear function, but most authors instead call the resulting class ESPACE.) If we use a nondeterministic machine instead, we get the class NEXPSPACE, which is equal to EXPSPACE by Savitch's theorem.

7 In logic, Linear temporal logic (LTL) is a modal temporal logic with modalities referring to time. In LTL, one can encode formulae about the future of paths such as that a condition will eventually be true, that a condition will be true until another fact becomes true, etc. It is a fragment of the more complex CTL*, which also allows branching time and quantifiers. Subsequently LTL is sometimes called propositional temporal logic, abbreviated PTL.

8 In proof theory, the semantic tableau (or truth tree) is a decision procedure for sentential and related logics, and a proof procedure for formulas of first-order logic. The tableau method can also determine the satisfiability of finite sets of formulas of various logics. It is the most popular proof procedure for modal logics (Girle 2000).

[9] In combinatorics and computer science, covering problems are computational problems that ask whether a certain combinatorial structure 'covers' another, or how large the structure has to be to do that. Covering problems are minimization problems and usually linear programs, whose dual problems are called packing problems.

[10] Control theory is an interdisciplinary branch of engineering and mathematics that deals with the behavior of dynamical systems. The desired output of a system is called the reference. When one or more output variables of a system need to follow a certain reference over time, a controller manipulates the inputs to a system to obtain the desired effect on the output of the system.

[11] In theoretical computer science, automata theory is the study of abstract machines (or more appropriately, abstract 'mathematical' machines or systems) and the computational problems that can be solved using these machines. These abstract machines are called automata. Automata comes from the Greek word αὐτόματα meaning "self-acting". The figure at right illustrates a finite state machine, which is one well-known variety of automaton. This automaton consists of states (represented in the figure by circles), and transitions (represented by arrows). As the automaton sees a symbol of input, it makes a transition (or jump) to another state, according to its transition function (which takes the current state and the recent symbol as its inputs).

[12] In computability theory, a system of data-manipulation rules (such as an instruction set, a programming language, or a cellular automaton) is said to be Turing complete or computationally universal if and only if it can be used to simulate any single-taped Turing machine and thus in principle any computer. A classic example is the lambda calculus. In practice Turing completeness, named after Alan Turing, means that the rules followed in sequence on arbitrary data can produce the result of any calculation. This requires, at a minimum, conditional branching (an "if" and "goto" statement) and the ability to change arbitrary memory locations (formality requires an explicit HALT state). To show that something is Turing complete, it is enough to show that it can be used to simulate the most primitive computer, since even the simplest computer can be used to simulate the most complicated one. All general purpose programming languages and modern machine instruction sets are Turing complete, notwithstanding finite-memory issues.

[13] A finite-state machine (FSM) or finite-state automaton (plural: automata), or simply a state machine, is a behavioral model used to design computer programs. It is composed of a finite number of states associated to transitions. A transition is a set of actions that starts from one state and ends in another (or the same) state. A transition is started by a trigger, and a trigger can be an event or a condition.

[14] A Markov chain, named for Andrey Markov, is a mathematical system that undergoes transitions from one state to another (from a finite or countable number of possible states) in a chainlike manner. It is a random process characterized as memoryless: the next state depends only on the current state and not on the entire past. This specific kind of "memorylessness" is called the Markov property. Markov chains have many applications as statistical models of real-world processes.

Chapter 15. State Transition Modeling

State transition modeling makes use of state transition diagrams to describe system behavior. These state transition diagrams use distinct states to define system behavior and changes. Most current modeling tools contain some kind of ability to represent state transition modeling. The use of state transition models can be most easily recognized as logic state diagrams and directed graphs for finite state machines.

State diagram

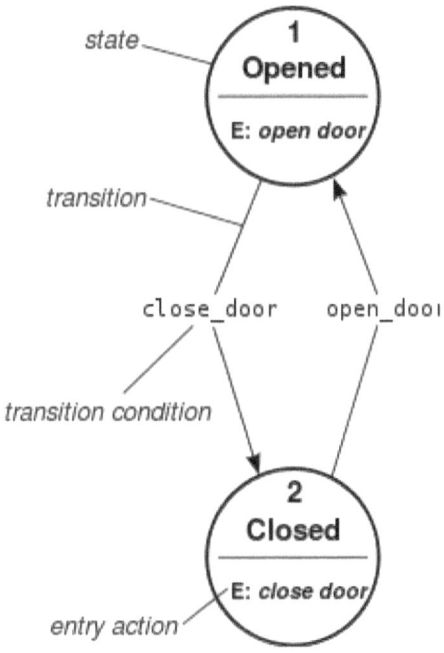

Figure 15-1. A state diagram for a door that can only be opened and closed

A **state diagram** is a type of diagram used in computer science and related fields to describe the behavior of systems. State diagrams require that the system described be composed of a finite number of states; sometimes, this is indeed the case, while at other times this is a reasonable abstraction. There are many forms of state diagrams, which differ slightly and have different semantics[1].

Overview

We use state diagrams to give an abstract description of the behavior of a system. This behavior is analyzed and represented in series of events that could occur in one or more possible states. Hereby *"each diagram usually represents objects of a single class and tracks the different states of its objects through the system"* [136].

State diagrams can be used to represent graphically finite state machines. Taylor Booth[2] introduced this in his 1967 book "Sequential Machines and Automata Theory". Another possible representation is the State transition table[3].

Directed graph

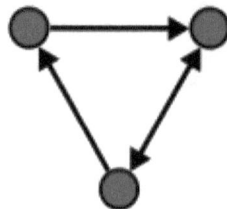

Figure 15-2. A directed graph

A classic form of state diagram for a finite state machine[4] is a directed graph with the following elements $(Q, \Sigma, Z, \delta, q_0, F)$: [137] [138]

Vertices Q: a finite set of states normally represented by circles and labeled with unique designator symbols or words written inside them.

Input symbols Σ: a finite collection of input symbols or designators.

Output symbols Z: a finite collection of output symbols or designators.

The output function ω represents the mapping of ordered pairs of input symbols and states onto output symbols, denoted mathematically as $\omega : \Sigma \times Q \to Z$.

Edges δ: represent the "transitions" between two states as caused by the input (identified by their symbols drawn on the "edges"). An edge is usually drawn as an arrow directed from the present-state toward the next-state. This mapping describes the state transition that is to occur on input of a particular symbol. This is written mathematically as $\delta : Q \times \Sigma \to Q$, so by δ (transition function) in definition of the FA is given both the pair of vertices connected by an edge and the symbol on an edge in a diagram representing this FA. Item $\delta(q, a) = p$ in definition of FA means that from the state named *q* occurs under input symbol **a** the transition to the state *p* in this machine. For the diagram representing this FA, this says that from the vertex labeled by *q* there is an edge labeled by **a** to the vertex labeled by *p*.

Start state q_0: (not shown in the examples below). The start state $q_0 \in Q$ is usually represented by an arrow with no origin pointing to the state. In older texts [137] [139], the start state is not shown and must be inferred from the text.

Accepting state(s) F: If used, for example for accepting automata, $F \in Q$ is the accepting state. It is usually drawn as a double circle. Sometimes the accept state(s) function as "Final" (halt, trapped) states [138].

For a deterministic finite state machine (DFA), nondeterministic finite state machine (NFA), generalized nondeterministic finite state machine (GNFA), or Moore machine, the input is denoted on each edge. For a Mealy machine, input and output are signified on each edge, separated with a slash "/": "1/0" denotes the state change upon encountering the symbol "1" causing the symbol "0" to be output. For a Moore machine, the state's output is usually written inside the state's circle, also separated from the state's designator with a slash "/". There are also variants that combine these two notations.

> Example: If a state has a number of outputs (e.g. "a = motor counter − clockwise = 1, b = caution light inactive = 0") the diagram should reflect this : e.g. "$q_5/1,0$" designates state q_5 with outputs $a = 1$, $b = 0$. This designator will be written inside the state's circle.

Example: DFA, NFA, GNFA, or Moore machine

S_1 and S_2 are states and S_1 is an **accepting state** or a **final state**. Each edge is labeled with the input. This example shows an acceptor for strings over {0,1} that contain an even number of zeros.

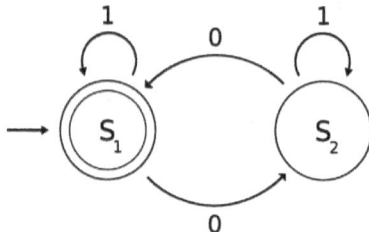

Figure 15-3. Example Moore machine

Example: Mealy machine

S_0, S_1, and S_2 are states. Each edge is labeled with "$j\ /\ k$" where j is the input and k is the output.

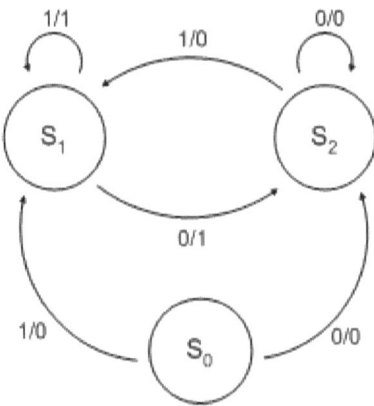

Figure 15-4. Example Mealy machine

Harel statechart

Harel[5] statecharts [140] are gaining widespread usage since a variant has become part of the Unified Modeling Language (UML). The diagram type allows the modeling of superstates, orthogonal regions, and activities as part of a state.

Classic state diagrams require the creation of distinct nodes for every valid combination of parameters that define the state. This can lead to a very large number of nodes and transitions between nodes for all but the simplest of systems (state and transition explosion). This complexity reduces the readability of the state diagram. With Harel statecharts, it is possible to model multiple cross-functional state diagrams within the state chart. Each of these cross-functional state machines can transition internally without affecting the other state machines in the statechart. The current state of each cross-functional state machine in the state chart defines the state of the system. The Harel state chart is equivalent to a state diagram but it improves the readability of the resulting diagram.

Alternative semantics

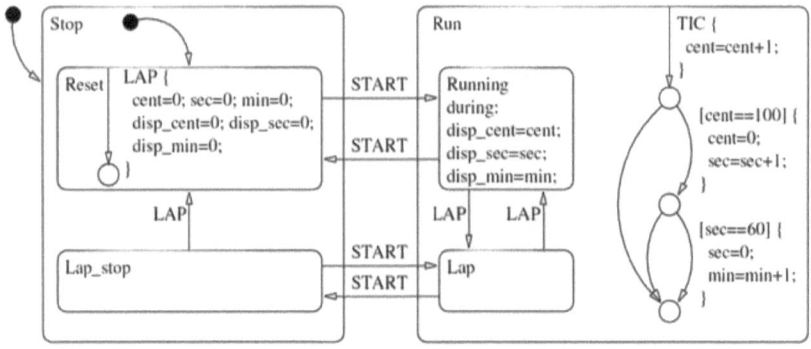

Figure 15-5. Model of a simple stopwatch [141]

There are other sets of semantics available to represent state diagrams. For example, there are tools for modeling and designing logic for embedded controllers [142]. They combine hierarchical state diagrams, flow graphs, and truth tables into one language, resulting in a different formalism and set of semantics [143]. The figure on the right illustrates this mix of state diagrams and flow graphs with a set of states to represent the state of a stopwatch and a flow graph to control the ticks of the watch. These diagrams, like Harel's original state machines [144], support hierarchically nested states, orthogonal regions, state actions, and transition actions [145].

State diagrams versus flowcharts

Newcomers to the state machine formalism often confuse **state diagrams** with **flowcharts**. The figure below shows a comparison of a state diagram with a flowchart. A state machine (panel (a)) performs actions in response to explicit events. In contrast, the flowchart (panel (b)) does not need explicit events but rather transitions from node to node in its graph automatically upon completion of activities [146].

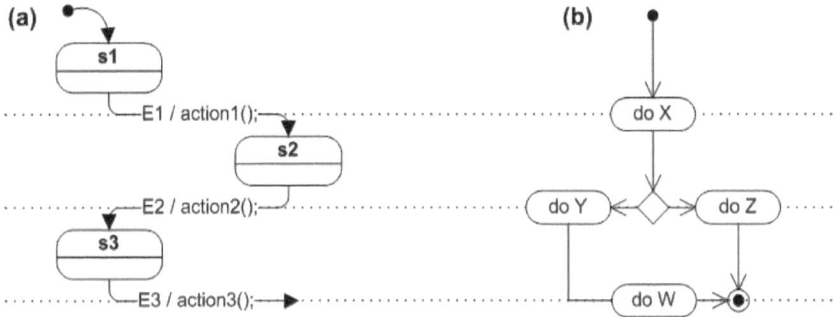

Figure 15-6. State diagram compared to flowchart

Graphically, compared to state diagrams, flowcharts reverse the sense of vertices and arcs. In a state diagram, the processing is associated with the arcs (transitions), whereas in a flowchart, it is associated with the vertices. A state machine is idle when it sits in a state waiting for an event to occur. A flowchart is busy executing activities when it sits in a node. The figure above attempts to show that reversal of roles by aligning the arcs of the state diagrams with the processing stages of the flowchart.

You can compare a flowchart to an assembly line in manufacturing because the flowchart describes the progression of some task from beginning to end (e.g., transforming source code input into object code output by a compiler). A state machine generally has no notion of such a progression. The door state machine shown at the top of this article, for example, is not in a more advanced stage when it is in the "closed" state, compared to being in the "opened" state; it simply reacts differently to the open/close events. A state in a state machine is an efficient way of specifying a particular behavior, rather than a stage of processing.

Other extensions

An interesting extension is to allow arcs to flow from any number of states to any number of states. This only makes sense if the system is allowed to be in multiple states at once, which implies that an individual state only describes a condition or other partial aspect of the overall, global state. The resulting formalism is known as a Petri net.

Another extension allows the integration of flowcharts within Harel state charts. This extension supports the development of software that is both event driven and workflow driven.

Notes

[1] Computational semantics is focused on the processing of linguistic meaning. In order to do this concrete algorithms and architectures are described. Within this framework, the algorithms and architectures are also analyzed in terms of decidability, time/space complexity, data structures that they require and communication protocols.

[2] Taylor L. Booth (September 22, 1933 - October 20, 1986) was a mathematician known for his work in automata theory. One of his fundamental works is Sequential Machines and Automata Theory (1967). It is a wide-ranging book meant for specialists, written for both theoretical computer scientists as well as electrical engineers. It deals with state minimization techniques, Finite state machines, Turing machines, Markov processes, and undecidability.

[3] In automata theory and sequential logic, a state transition table is a table showing what state (or states in the case of a nondeterministic finite automaton) a finite semiautomaton or finite state machine will move to, based on the current state and other inputs. A state table is essentially a truth table in which some of the inputs are the current state, and the outputs include the next state, along with other outputs. A state table is one of many ways to specify a state machine, other ways being a state diagram, and a characteristic equation.

[4] A finite-state machine (FSM) or finite-state automaton (plural: automata), or simply a state machine, is a behavioral model used to design computer programs. It is composed of a finite number of states associated to transitions. A transition is a set of actions that starts from one state and ends in another (or the same) state. A transition is started by a trigger, and a trigger can be an event or a condition.

[5] David Harel (born 1950) is a professor of computer science at the Weizmann Institute of Science in Israel. Born in London, England, he was

Dean of the Faculty of Mathematics and Computer Science at the institute for seven years. Harel is best known for his work on dynamic logic, computability and software engineering. In the 1980s, he invented the graphical language of Statecharts, which has been adopted as part of the UML standard.

Chapter 16. Object-Role Modeling

Object Role Modeling (ORM) is a method for conceptual modeling, and can be used as a tool for information and rules analysis, ontological analysis, and data modeling in the field of software engineering [147].

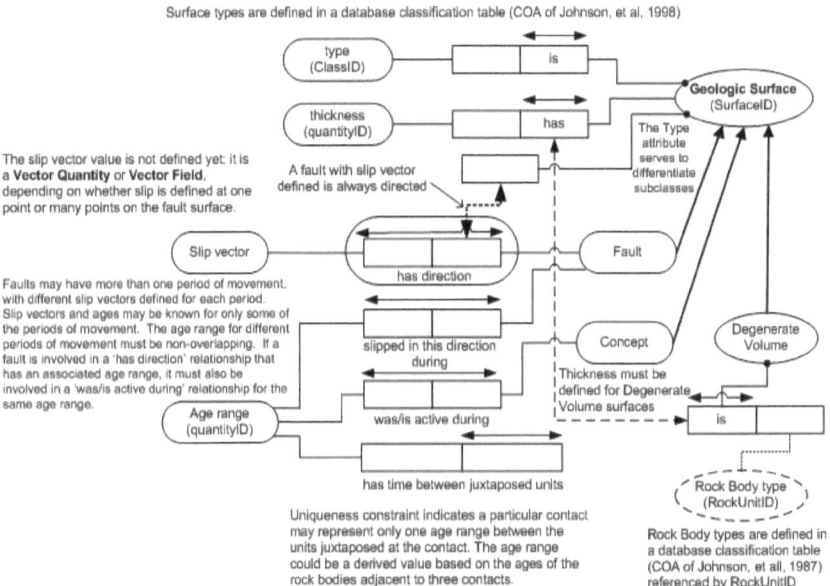

Figure 16-1. Example of application of Object Role Modeling in "Schema for Geologic Surface", Stephen M. Richard (1999) [148].

Overview

Object Role Modeling is a fact-oriented method for performing systems analysis at the conceptual level. The quality of a database application depends critically on its design. To help ensure correctness, clarity, adaptability and productivity, information systems are best specified first at the conceptual level, using concepts and language that people can readily understand. The conceptual design may include data, process and behavioral perspectives, and the actual DBMS used to implement the design might be based on one of many logical data models (relational, hierarchic, network, object-oriented etc.) [18].

The designer of a database builds a formal model of the application area or Universe of Discourse[1] (UoD). The model requires a good understanding of the UoD and a means of specifying this understanding in a clear, unambiguous way. Object-Role Modeling (ORM) simplifies the design process by using natural language, as well as intuitive diagrams that can be populated with examples, and by examining the information in terms of simple or elementary facts. By expressing the model in terms of natural concepts, like objects and roles, it provides a conceptual approach to modeling. Its attribute-free approach promotes semantic stability [149].

History

The roots of ORM can be traced to research into semantic modeling for information systems in Europe during the 1970s. There were many pioneers and this short summary does not mention them all by any means. An early contribution came in 1973 when Michael Senko wrote about "*data structuring*" in the IBM Systems Journal. In 1974, Jean-Raymond Abrial contributed an article about "Data Semantics". In June 1975, Eckhard Falkenberg's doctoral thesis was published and in 1976, one of Falkenberg's papers mentions the term "object-role model".

G. M. Nijssen[2] made fundamental contributions by introducing the "circle-box" notation for object types and roles, and by formulating the first version of the conceptual schema design procedure. Robert Meersman extended the approach by adding sub-typing, and introducing the first truly conceptual query language.

Object role modeling also evolved from the *Natural language Information Analysis Method*, a methodology that was initially developed by the academic researcher, G. M. Nijssen in the Netherlands (Europe) in the mid-1970s and his research team at the Control Data Corporation Research Laboratory in Belgium, and later at the University of Queensland, Australia in the 1980s. The acronym **NIAM** originally stood for "Nijssen's Information Analysis Methodology", and later generalized to "Natural language Information Analysis Methodology" and *Binary Relationship Modeling* since G. M. Nijssen was only one of many people involved in the development of the method.

In 1989, Terry Halpin[3] completed his PhD thesis on ORM, providing the first full formalization of the approach and incorporating several extensions.

Also in 1989, Terry Halpin and G. M. Nijssen co-authored the book "Conceptual Schema and Relational Database Design" and several joint papers, providing the first formalization of Object-Role Modeling. Since then Dr. Terry Halpin has authored six books and over 160 technical papers.

A recent variation of ORM is referred to as FCO-IM. It distinguishes itself from traditional ORM in that it takes a strict communication-oriented perspective. Rather than modeling the domain and its essential concepts, it purely models the grammar used to discourse about the domain. Another recent development is the use of ORM in combination with standardized relation types with associated roles and a standard machine-readable dictionary and taxonomy of concepts as are provided in the Gellish English dictionary. Standardization of relation types (fact types), roles and concepts enables increased possibilities for model integration and model reuse.

ORM2

ORM2 (second-generation ORM) is a new incarnation of Object-Role Modeling, supported by various modeling tools to support the new notation. Dr. Terry Halpin lead the effort of the NORMA ORM modeling tool spearheaded by Neumont University and currently through The ORM Foundation [150], a UK-based non-profit organization dedicated to the promotion of the fact-oriented approach to information modeling

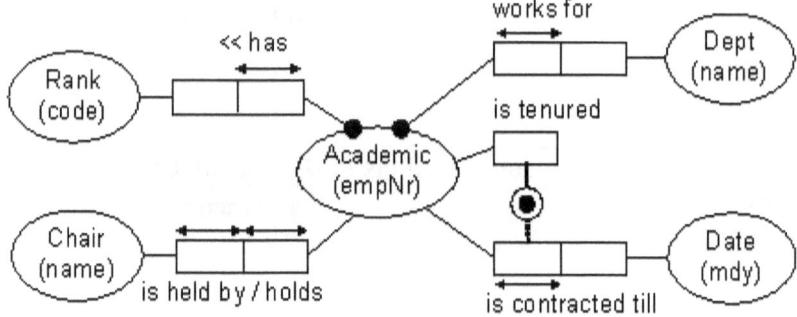

Figure 16-2. Example of an ORM2 diagram

ORM2 leverages the work done by the BSBR group.

The main objectives for the ORM 2 graphical notation are [151]:

- More compact display of ORM models without compromising clarity
- Improved internationalization (e.g. avoid English language symbols)
- Notation changes acceptable to a short-list of key ORM users
- Simplified drawing rules to facilitate creation of a graphical editor
- Full support of textual annotations (e.g. footnoting of textual rules)
- Extended use of views for selectively displaying/suppressing detail
- Support for new features (e.g. role path delineation, closure aspects, modalities)

The NORMA (Natural ORM Architect) tool is an open source project incorporating the ORM2 syntax.

Object role modeling topics
Graphic notation

ORM's rich graphic notation is capable of capturing many business rules that are typically unsupported as graphic primitives in other popular data modeling notations.

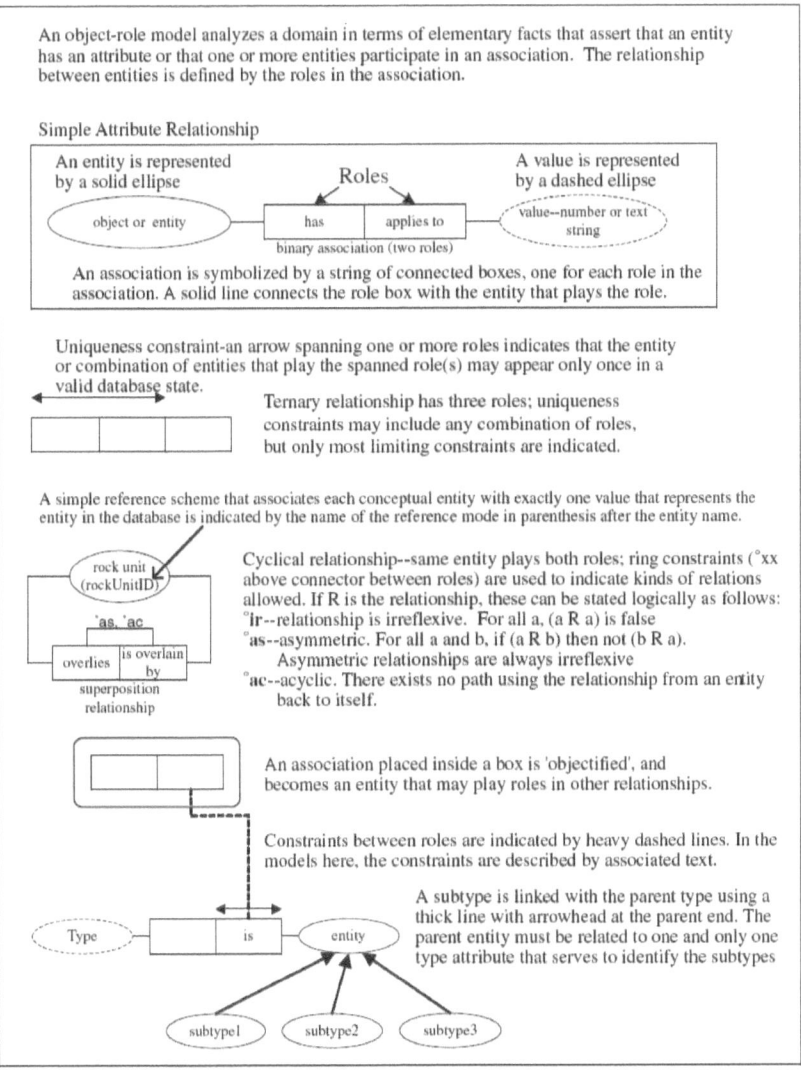

Figure 16-3. Overview of the Object-Role Model notation, Stephen M. Richard (1999) [148].

Various software tools exist to enter ORM schemas, and generate relational database schemas. These include Microsoft Visio for Enterprise Architects, OORIANE, CaseTalk, Infagon, and NORMA.

A graphical NIAM design tool that included the ability to generate database-creation scripts for Oracle, DB2 and DBQ was developed in the early 1990s in Paris. It was originally named Genesys and was marketed successfully in France and later Canada. It could also handle ER diagram design. It was ported to SCO Unix, SunOs, DEC 3151's and Windows 3.0 platforms, and was later migrated to succeeding Microsoft operating systems, utilizing XVT for cross operating system graphical portability. The tool was renamed OORIANE and is currently being used for large data warehouse and SOA projects.

The conceptual schema design procedure

The information system's life cycle typically involves several stages: feasibility study; requirements analysis; conceptual design of data and operations; logical design; external design; prototyping; internal design and implementation; testing and validation; and maintenance. ORM's conceptual schema design procedure (CSDP) focuses on the analysis and design of data. The seven steps of the conceptual schema design procedure [18]:

1. Transform familiar information examples into elementary facts, and apply quality checks
2. Draw the fact types, and apply a population check
3. Check for entity types that should be combined, and note any arithmetic derivations
4. Add uniqueness constraints, and check arity[4] of fact types
5. Add mandatory role constraints, and check for logical derivations
6. Add value, set comparison and sub-typing constraints
7. Add other constraints and perform final checks

Tools

The growth of ORM has followed the availability of a series of steadily improving ORM tools.

The early ORM tools such as IAST (Control Data) and RIDL* were followed by InfoDesigner, InfoModeler and VisioModeler.

When Microsoft bought the Visio Corporation, Microsoft extended VisioModeler and made it a component of Microsoft Visual Studio. This was Microsoft's first ORM implementation and it was published in the 2003 Enterprise Architects release of Visual Studio as a component of the tool called "Microsoft Visio for Enterprise Architects (VEA)".

In the same year, a companion "how to" book was published by Morgan Kaufmann entitled "Database Modeling with Microsoft Visio for Enterprise Architects" [152]. Microsoft has retained the ORM functionality in the high-end version of Visual Studio 2005 and the Morgan Kaufmann book remains a suitable user guide for both versions.

The next "in the works" ORM tool is an open source tool called NORMA (Natural ORM Architect for Visual Studio).

DogmaModeler

DogmaModeler is a free ontology-modeling tool based on Object role modeling. The philosophy of DogmaModeler is to enable non-IT experts to model ontologies with a little or no involvement of an ontology engineer.

This challenge is tackled in DogmaModeler through well-defined methodological principles. The first version of DogmaModeler was developed at the Vrije Universiteit Brussels.

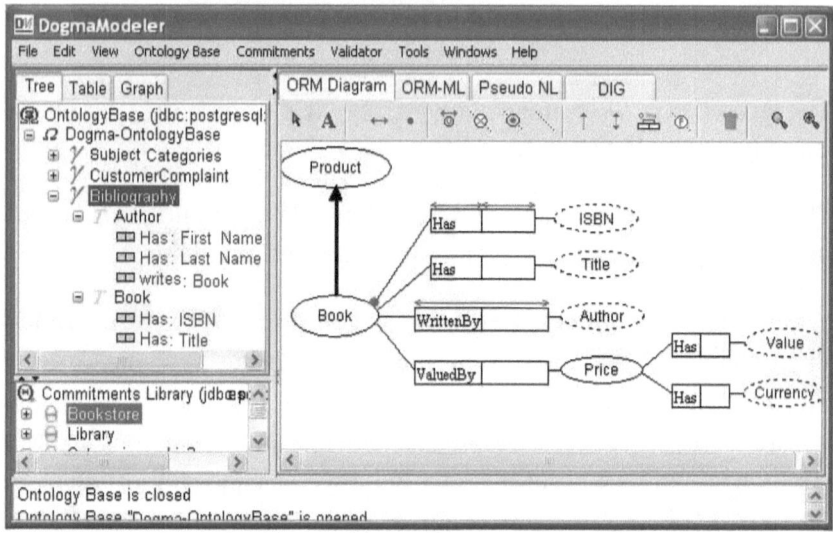

Figure 16-4. DogmaModeler Screenshot [153].

DogmaModeler open source status

The DogmaModeler Project [154] shows no activity since its creation in 2006, and the source code for the project is not available through that site. The latest version of the program, available at the http://jarrar.info/Dogmamodeler website is dated on October 27, 2006.

Since then the project seems to have been continued and expanded into several other tools at the Vrije Universiteit Brussel's Semantics Technology and Applications Research Laboratory (VUB STARLab) [155]. A note on that site states "Users who only use DogmaModeler for their own researches can contact (author's email) for the free download" [156].

VisioModeler

The former ORM tool known as VisioModeler is freely available as an unsupported product from Microsoft Corporation (as a 25 MB download). Models developed in VisioModeler may be exported to Microsoft's current and future ORM solutions. To obtain the free VisioModeler download, go to http://download.microsoft.com, search by selecting Keyword Search, enter the keyword "VisioModeler", select your operating system (e.g. Windows XP— Note: VisioModeler does NOT work under Windows Vista), change the setting for "Show Results for" to "All Downloads", and hit the "Find It!" button. This should bring up a download page that includes the title "VisioModeler (Unsupported Product Edition)". Clicking on this will take you to the link for the download file MSVM31.exe. Click on this to do the download. I have tested this on Windows 7 and it does not run. However, the ORM foundation has developed a work-around (see http://www.ormfoundation.org/ files/folders/visomodeler/category1092.aspx)

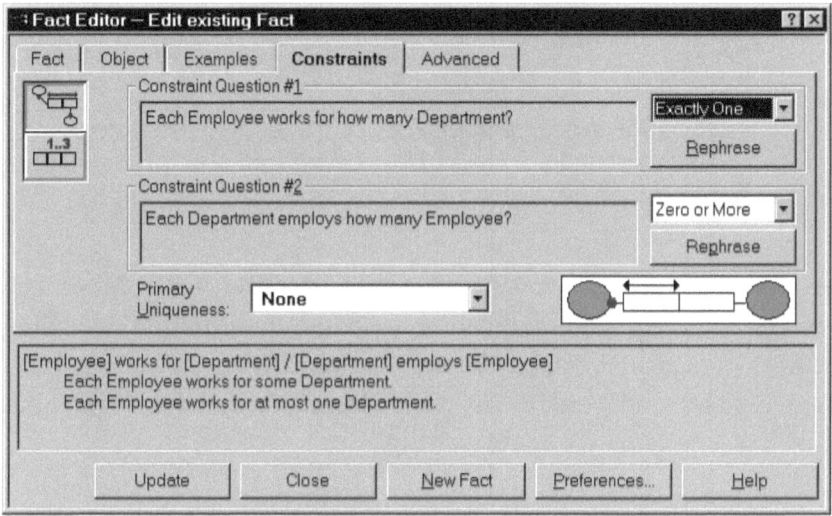

Figure 16-5. Screenshot of VisioModeler

Visio for Enterprise Architects (VEA)

Microsoft included a powerful ORM and logical database modeling solution within its Visio for Enterprise Architects (VEA) product. The 2005 release of VEA also included some minor upgrades (e.g., a driver for SQL Server 2005 was included).

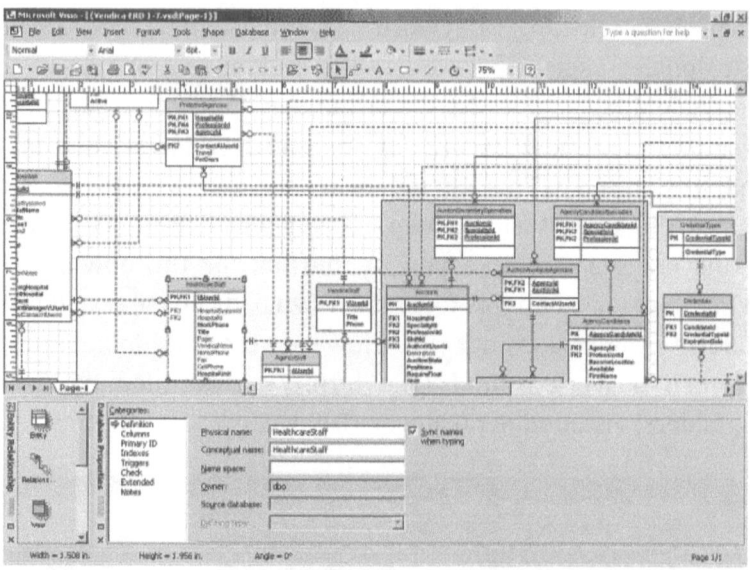

Figure 16-6. Screenshot of Viso for Enterprise Architects

CaseTalk

A modeling tool called CaseTalk [157] based on the ORM-dialect known as Fully Communication Oriented Information Modeling (FCO-IM) is available from Bommeljé Crompvoets en partners b.v., headquartered in Utrecht, The Netherlands. To find out more about this tool, check the CaseTalk website.

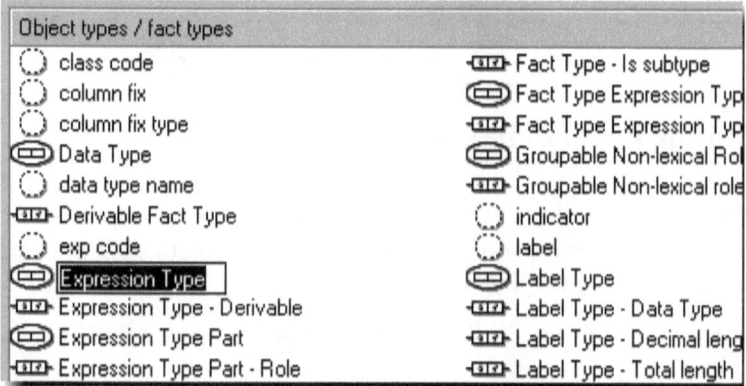

Figure 16-7. The Object Types/Fact Types in the left part of the screen is now expanded with a multi-columned listview.

Infagon

A freeware ORM tool known as Infagon is available from Mattic software. Infagon is also based on the FCO-IM dialect. To download or obtain more details about this tool, click the Infagon home page [158].

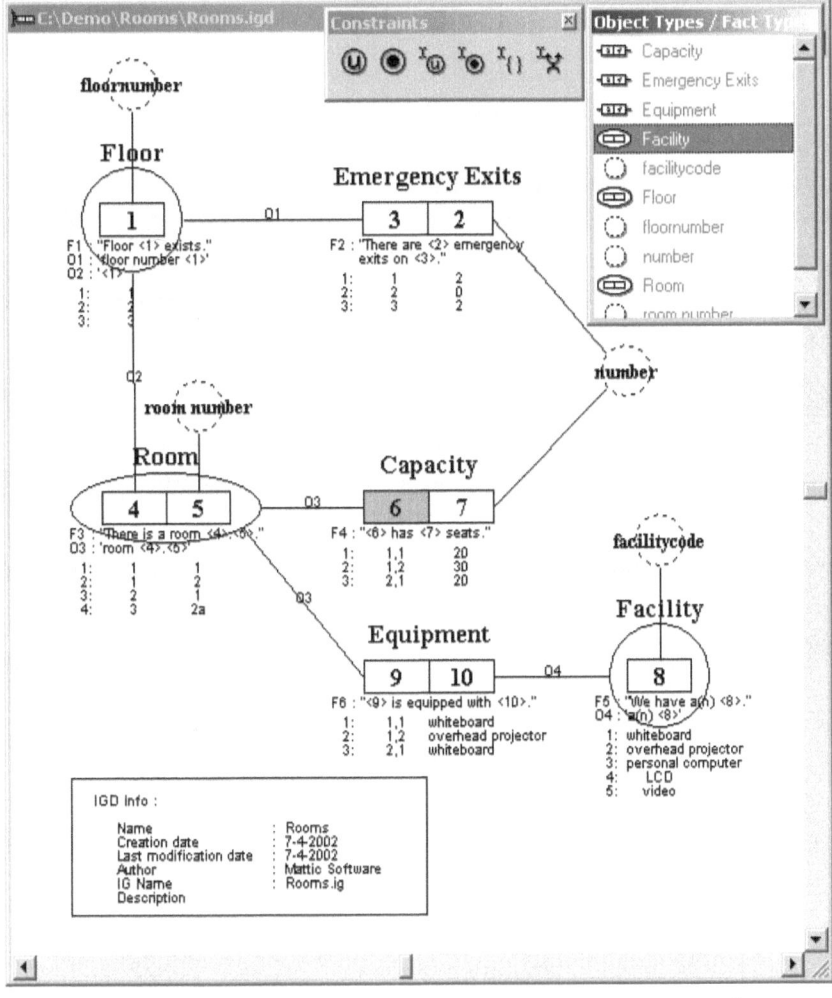

Figure 16-8. Infagon's Drag-and-drop Diagramming interface

Other tools

- **Doctool** and **CogNIAM** (CogNIAM tools) [159].
- **ActiveFacts** (ORM 2 tools) [160],
- **DogmaStudio** (ORM Ontology tool) [161],
- **Orthogonal Toolbox** (free XML add-on to database modeling COM API for Microsoft's ORM solution) [162].

NORMA

NORMA (Natural ORM Architect for Visual Studio) [150] is a free and open source plug-in to Microsoft Visual Studio 2005, Visual Studio 2008 and Visual Studio 2010. It supports ORM 2 (second generation ORM), and maps ORM models to a variety of implementation targets, including major database engines, object-oriented code, and XML schema.

Dr. Terry Halpin's latest book, Information Modeling and Relational Databases, Second Edition [163] "...*uses the notation of ORM 2 (second generation ORM), as supported by the NORMA (Neumont ORM Architect) tool...*" (page 10), and "*(...) At the time of writing, the Neumont ORM Architect (NORMA) tool provides the most complete support for the ORM 2 notation discussed in this book.*" (Preface, page xxv).

NORMA Project status

As of April 2009, the NORMA project [164] delivers frequent releases and is almost ready to encompass the entire life-cycle of databases.

NORMA supports multiple generation targets including:

| Database engines | Microsoft Sql Server, Oracle, DB2, MySQL, PostgreSQL, etc. |

Programming languages	LINQ to SQL, PLiX (Programming Language in XML) [165] and PHP
Other	XML schemas (XSD)

Advantages of NORMA are these:

- Accepts typed input and automatically generates graphics
- Validates common constraints and completeness as the model is entered
- Provides simultaneous narrative and graphic versions of all models
- Can provide E-R as well as ORM views of the model
- Automatic navigation from error message to graphic view of the error
- The three views (ORM, E-R, and narrative) provide comfortable access for most viewers. (Not everyone is comfortable with graphics, and the narrative view of the model is more quickly read by those who are comfortable with graphics.)
- Can reverse-engineer a physical database (up to a point)
- Narrative view uses hyperlinks for full cross-referencing
- Graphic model has no fixed bounds
- Automatic navigation from error message to graphic view of the error

Notes

[1] In the formal sciences, the domain of discourse, also called the universe of discourse (or simply universe), is the set of entities over which certain variables of interest in some formal treatment may range. The domain of discourse is usually identified in the preliminaries, so that there is no need in the further treatment to specify each time the range of the relevant variables.

[2] Sjir Nijssen (born 1938) is a Dutch computer scientist, who was fulltime professor at the University of Queensland. Nijssen is considered the founder of verbalization in computer science, and one of the founders of business modeling and information analysis based on natural language

[3] Terence Aidan (Terry) Halpin is an Australian computer scientist who is known for his formalization of the object role modeling notation.

[4] In logic, mathematics, and computer science, the arity i of a function or operation is the number of arguments or operands that the function takes.

Chapter 17. Unified Modeling Language

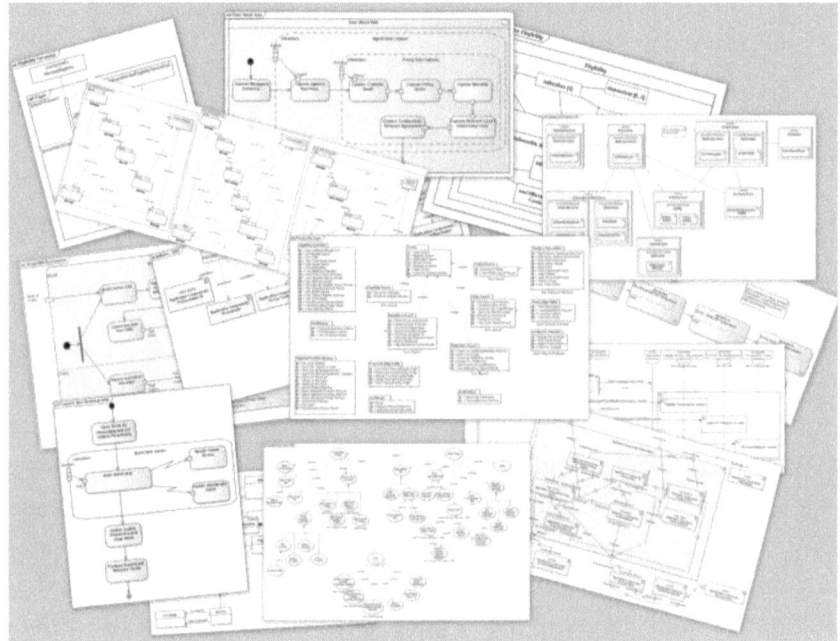

Figure 17-1. A collage of UML diagrams.

Unified Modeling Language (UML) is a standardized general-purpose modeling language in the field of object-oriented software engineering. The Object Management Group created and manages the standard.

UML includes a set of graphic notation techniques to create visual models of object-oriented software-intensive systems.

Overview

The Unified Modeling Language (UML) is used to specify, visualize, modify, construct and document the artifacts of an object-oriented software-intensive system under development [166]. UML offers a standard way to visualize a system's architectural blueprints, including elements such as:

- activities
- actors
- business processes
- database schemas
- (logical) components
- programming language statements
- reusable software components [167] [168] [169].

UML combines techniques from data modeling (entity relationship diagrams), business modeling (work flows), object modeling, and component modeling. It can be used with all processes, throughout the software development life cycle, and across different implementation technologies [170]. UML has synthesized the notations of the Booch method, the Object-modeling technique (OMT) and Object-oriented software engineering (OOSE) by fusing them into a single, common and widely usable modeling language. UML aims to be a standard modeling language, which can model concurrent and distributed systems. UML is a de facto industry standard,[1] and is evolving under the auspices of the Object Management Group (OMG).

UML models may be automatically transformed to other representations (e.g. Java) by means of QVT-like transformation languages. UML is extensible[1], with two mechanisms for customization: profiles[2] and stereotypes[3].

History
Before UML 1.x

After Rational Software Corporation hired James Rumbaugh[4] from General Electric in 1994, the company became the source for the two most popular object-oriented modeling approaches of the day: Rumbaugh's Object-modeling technique (OMT), which was better for object-oriented analysis[5] (OOA), and Grady Booch's[6] Booch method, which was better for object-oriented design (OOD). They were soon assisted in their efforts by Ivar Jacobson[7], the creator of the object-oriented software engineering (OOSE) method. Jacobson joined Rational in 1995, after his company, Objectory AB[8], was acquired by Rational. The three methodologists were collectively referred to as the *Three Amigos*.

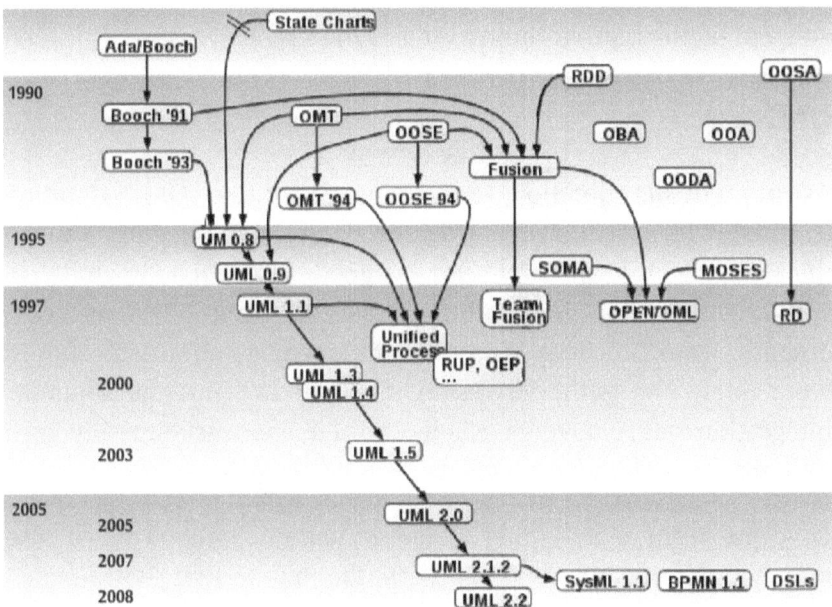

Figure 17-2. History of object-oriented methods and notation.

In 1996, Rational concluded that the abundance of modeling languages was slowing the adoption of object technology, so repositioning the work on a unified method; they tasked the Three Amigos with the development of a non-proprietary Unified Modeling Language. Representatives of competing object technology companies were consulted during OOPSLA[9] '96; they chose *boxes* for representing classes rather than the *cloud* symbols that were used in Booch's notation.

Under the technical leadership of the Three Amigos, an international consortium called the UML Partners was organized in 1996 to complete the *Unified Modeling Language (UML)* specification, and propose it as a response to the OMG RFP. The UML Partners' UML 1.0 specification draft was proposed to the OMG in January 1997. During the same month, the UML Partners formed a Semantics Task Force, chaired by Cris Kobryn and administered by Ed Eykholt, to finalize the semantics of the specification and integrate it with other standardization efforts. The result of this work, UML 1.1, was submitted to the OMG in August 1997 and adopted by the OMG in November 1997 [171].

UML 1.x

As a modeling notation, the influence of the OMT notation dominates (e. g., using rectangles for classes and objects). Though the Booch "cloud" notation was dropped, the Booch capability to specify lower-level design detail was embraced. The use case notation from Objectory and the component notation from Booch were integrated with the rest of the notation, but the semantic integration was relatively weak in UML 1.1, and was not really fixed until the UML 2.0 major revision.

Concepts from many other object oriented methods were also loosely integrated with UML with the intent that UML would support all object oriented methods. Many others also contributed, with their approaches flavoring the many models of the day, including: Tony Wasserman and Peter Pircher with the "Object-Oriented Structured Design (OOSD)" notation (not a method), Ray Buhr's "Systems Design with Ada", Archie Bowen's use case and timing analysis, Paul Ward's data analysis and David Harel's "Statecharts"; as the group tried to ensure broad coverage in the real-time systems domain. As a result, UML is useful in a variety of engineering problems, from single process, single user applications to concurrent, distributed systems, making UML rich but also large.

The Unified Modeling Language is an international standard: ISO/IEC 19501:2005 Information technology – Open Distributed Processing – Unified Modeling Language (UML) Version 1.4.2

UML 2.x

UML has matured significantly since UML 1.1. Several minor revisions (UML 1.3, 1.4, and 1.5) fixed shortcomings and bugs with the first version of UML, followed by the UML 2.0 major revision that was adopted by the OMG in 2005 [168].

Although OMG never released UML 2.1 as a formal specification, versions 2.1.1 and 2.1.2 appeared in 2007, followed by UML2.2 in February 2009. UML 2.3 was formally released in May 2010 [172]. UML 2.4 is in the beta stage as of March 2011 [173].

There are four parts to the UML 2.x specification:

1. The Superstructure that defines the notation and semantics for diagrams and their model elements
2. The Infrastructure that defines the core metamodel on which the Superstructure is based
3. The Object Constraint Language (OCL) for defining rules for model elements
4. The UML Diagram Interchange that defines how UML 2 diagram layouts are exchanged

The current versions of these standards follow: UML Superstructure version 2.3, UML Infrastructure version 2.3, OCL version 2.2, and UML Diagram Interchange version 1.0 [174].

Although many UML tools support some of the new features of UML 2.x, the OMG provides no test suite to test objectively compliance with its specifications.

Topics
Software development methods

UML is not a development method by itself [175]; however; it was designed to be compatible with the leading object-oriented software development methods of its time (for example OMT, Booch method,

Objectory). Since UML has evolved, some of these methods have been recast to take advantage of the new notations (for example OMT), and new methods have been created based on UML, such as IBM Rational Unified Process (RUP). Others include Abstraction Method and Dynamic Systems Development Method.

Modeling

It is important to distinguish between the UML model and the set of diagrams of a system. A diagram is a partial graphic representation of a system's model. The model also contains documentation that drives the model elements and diagrams (such as written use cases).

UML diagrams represent two different views of a system model [176]:

- Static view (or *structural view*): emphasizes the static structure of the system using objects, attributes, operations and relationships. The structural view includes class diagrams and composite structure diagrams.
- Dynamic view (or *behavioral view*): emphasizes the dynamic behavior of the system by showing collaborations among objects and changes to the internal states of objects. This view includes sequence diagrams, activity diagrams and state machine diagrams.

UML models can be exchanged among UML tools by using the XMI interchange format.

Diagrams overview

UML 2.2 has 14 types of diagrams divided into two categories [177]. Seven diagram types represent *structural* information, and the other seven represent general types of *behavior*, including four that represent different aspects of *interactions*. These diagrams can be categorized hierarchically as shown in the following class diagram:

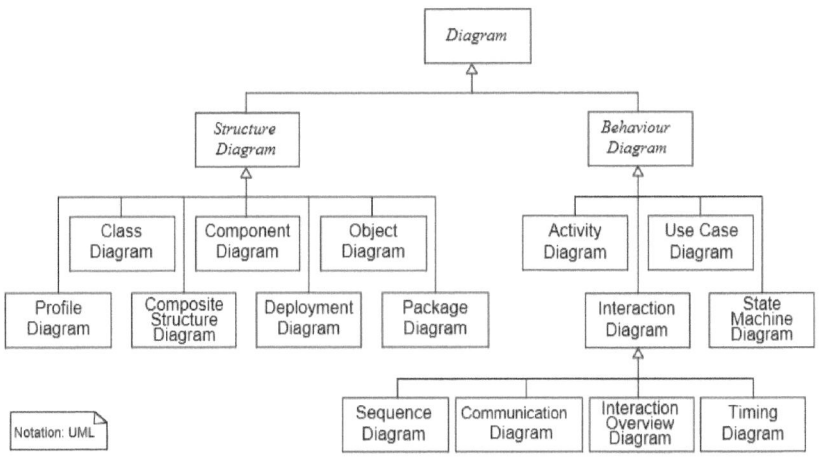

Figure 17-3. UML Diagram Overview

UML does not restrict UML element types to a certain diagram type. In general, every UML element may appear on almost all types of diagrams; this flexibility has been partially restricted in UML 2.0. UML profiles may define additional diagram types or extend existing diagrams with additional notations.

In keeping with the tradition of engineering drawings, a comment or note explaining usage, constraint, or intent is allowed in a UML diagram.

Structure diagrams

Structure diagrams emphasize the things that must be present in the modeled system. Since structure diagrams represent the structure, we may use them extensively in documenting the software architecture of software systems.

- Class diagram: describes the structure of a system by showing the system's classes, their attributes, and the relationships among the classes.
- Component diagram: describes how a software system is split up into components and shows the dependencies among these components.

- Composite structure diagram: describes the internal structure of a class and the collaborations that this structure makes possible.
- Deployment diagram: describes the hardware used in system implementations and the execution environments and artifacts deployed on the hardware.
- Object diagram: shows a complete or partial view of the structure of an example modeled system at a specific time.
- Package diagram: describes how a system is split up into logical groupings by showing the dependencies among these groupings.
- Profile diagram: operates at the metamodel level to show stereotypes as classes with the <<stereotype>> stereotype, and profiles as packages with the <<profile>> stereotype. The extension relation (solid line with closed, filled arrowhead) indicates what metamodel element a given stereotype is extending.

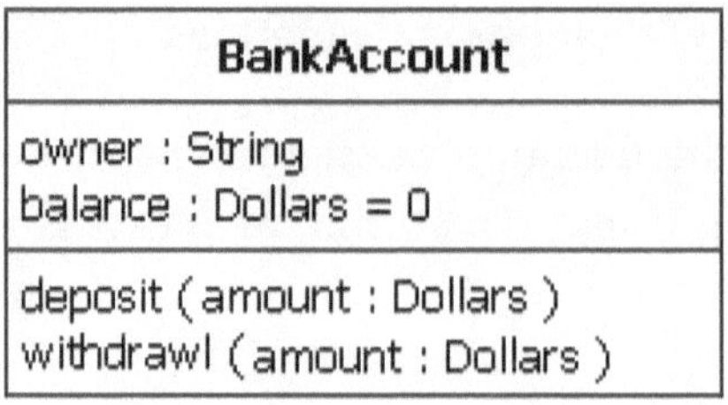

BankAccount
owner : String balance : Dollars = 0
deposit (amount : Dollars) withdrawl (amount : Dollars)

Figure 17-4. Class diagram

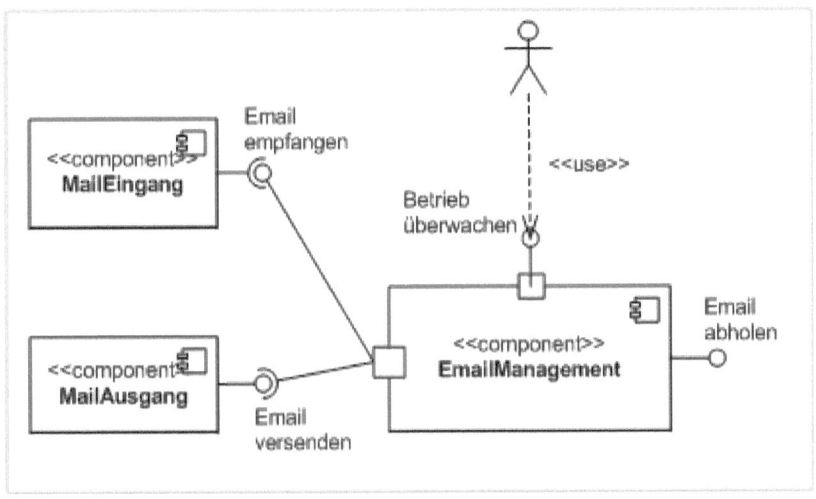

Figure 17-5. Component diagram

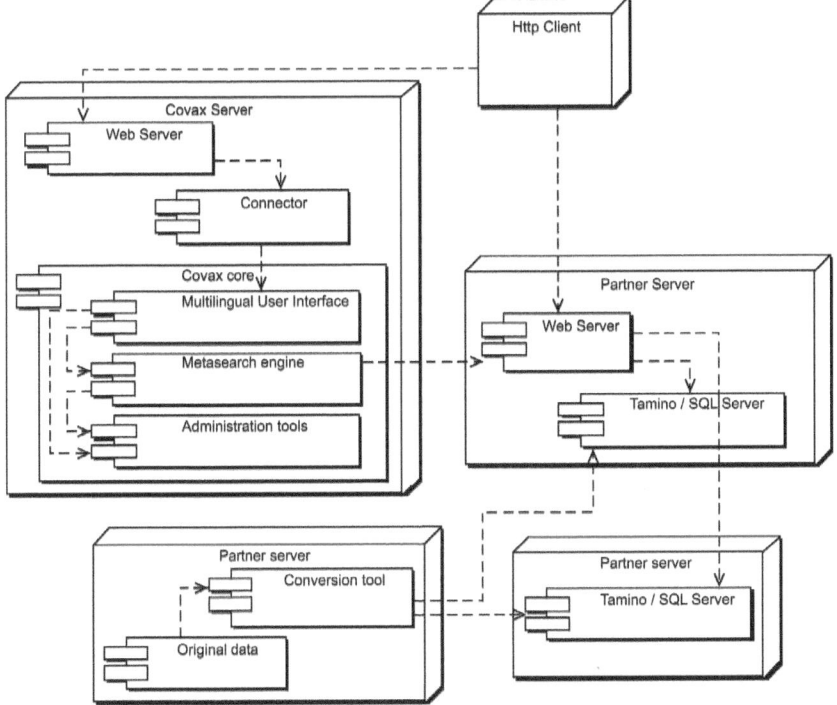

Figure 17-6. Deployment diagram

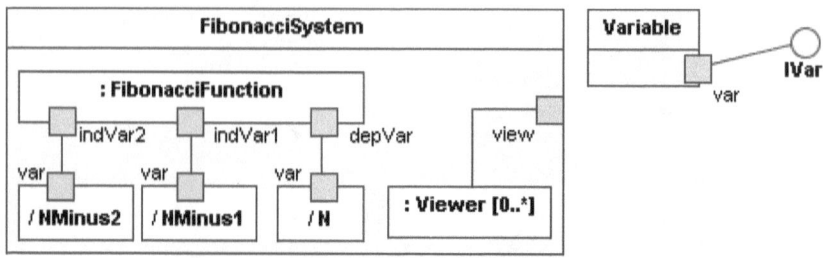

Figure 17-7. Composite structure diagrams

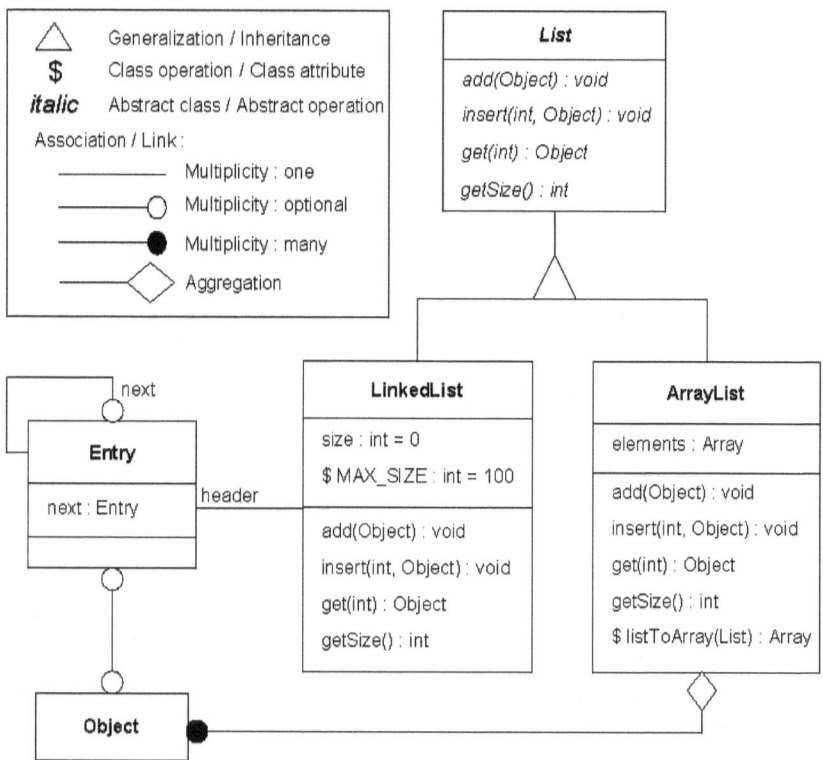

Figure 17-8. Object diagram

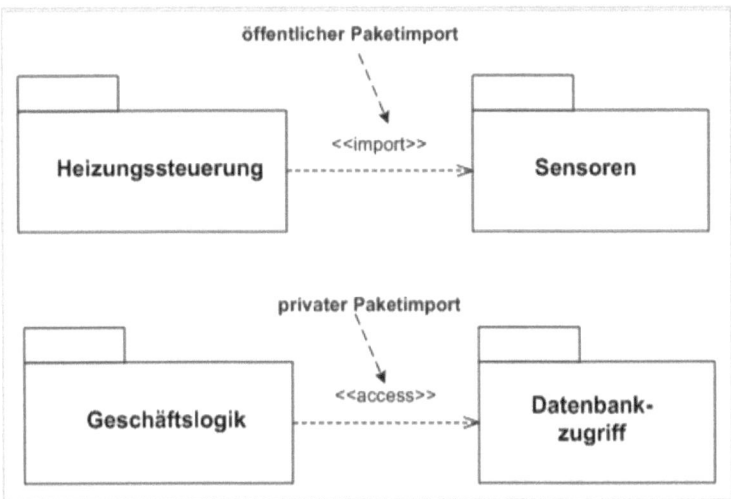

Figure 17-9. Package diagram

Behavior diagrams

Behavior diagrams emphasize what must happen in the system being modeled. Since behavior diagrams illustrate the behavior of a system, they are used extensively to describe the functionality of software systems.

- Activity diagram: describes the business and operational step-by-step workflows of components in a system. An activity diagram shows the overall flow of control.
- UML state machine diagram: describes the states and state transitions of the system.
- Use case diagram: describes the functionality provided by a system in terms of actors, their goals represented as use cases, and any dependencies among those use cases.

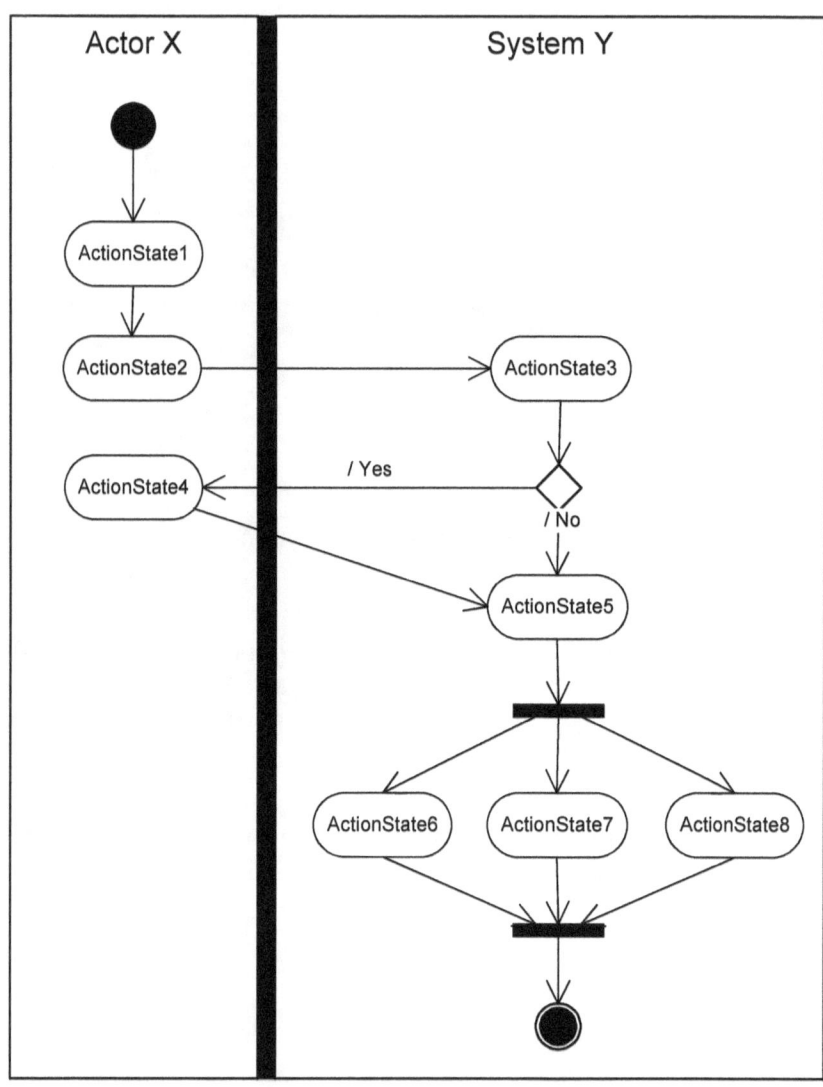

Figure 17-10. UML Activity Diagram

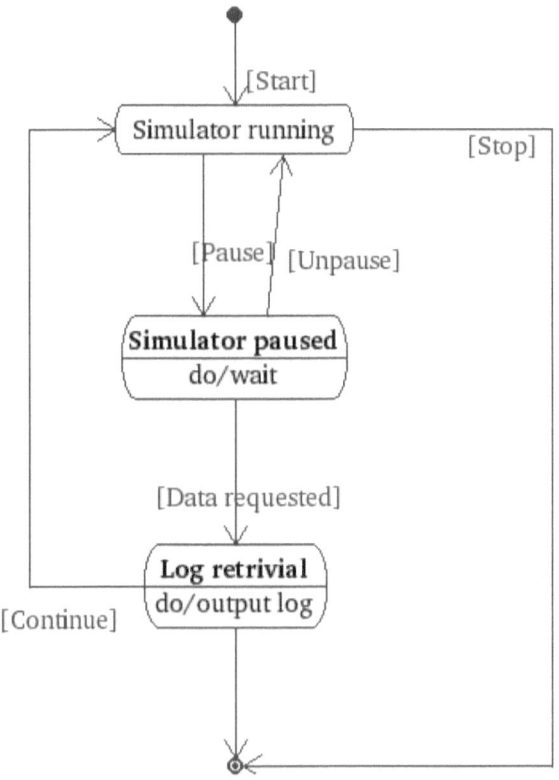

Figure 17-11. State Machine diagram

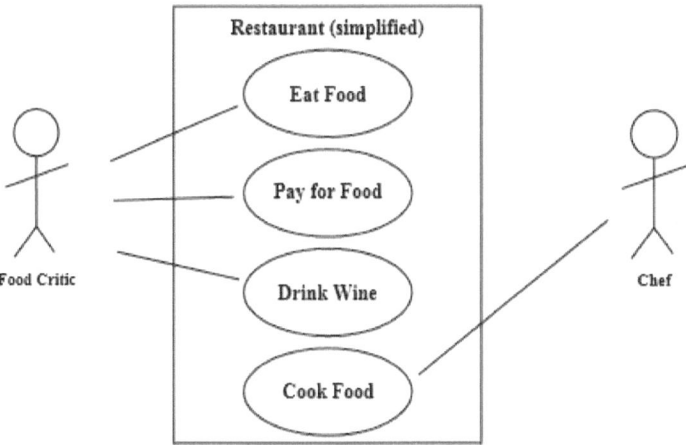

Figure 17-12. Use case diagram

Interaction diagrams

Interaction diagrams, a subset of behavior diagrams, emphasize the flow of control and data among the things in the system being modeled:

- Communication diagram: shows the interactions between objects or parts in terms of sequenced messages. They represent a combination of information taken from Class, Sequence, and Use Case Diagrams describing both the static structure and dynamic behavior of a system.
- Interaction overview diagram: provides an overview in which the nodes represent communication diagrams.
- Sequence diagram: shows how objects communicate with each other in terms of a sequence of messages. Also indicates the lifespans of objects relative to those messages.
- Timing diagrams: a specific type of interaction diagram where the focus is on timing constraints.

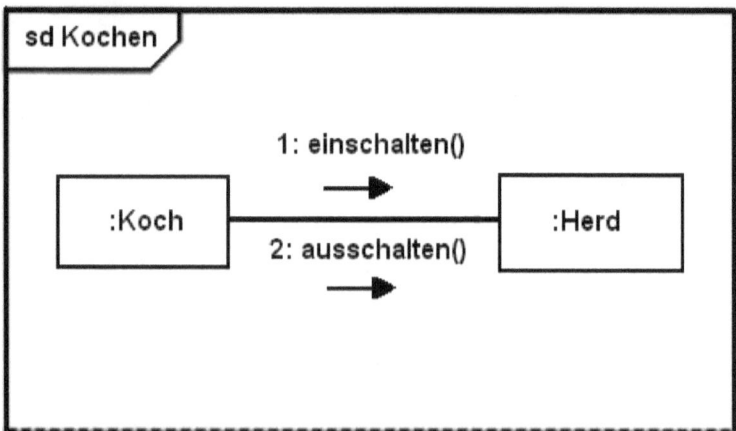

Figure 17-13. Communication diagram

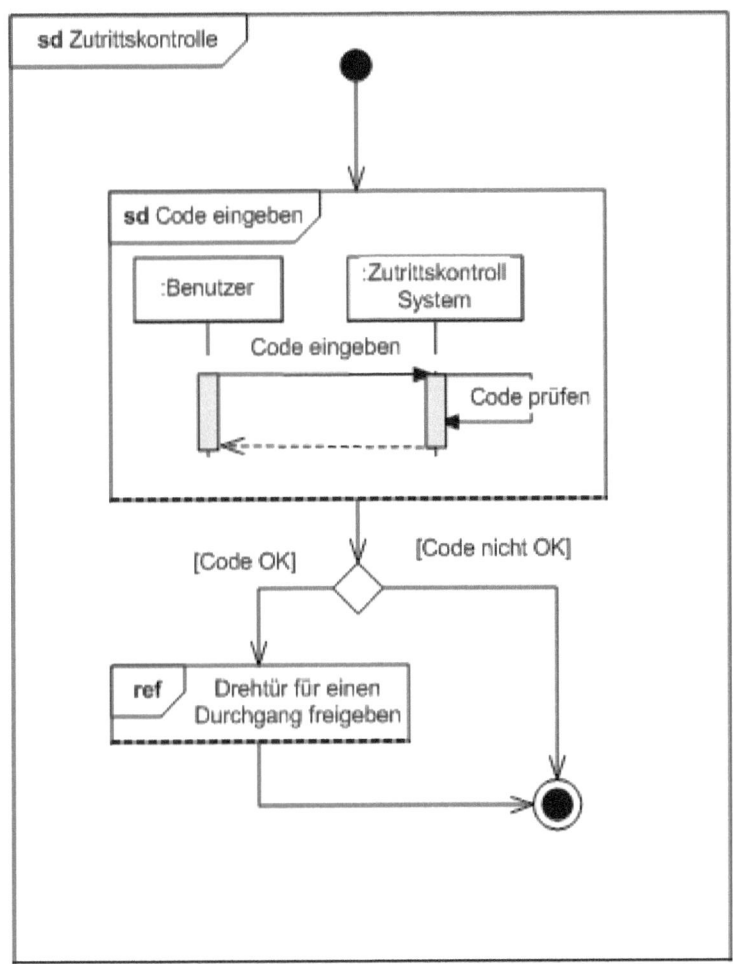

Figure 17-14. Interaction overview diagram

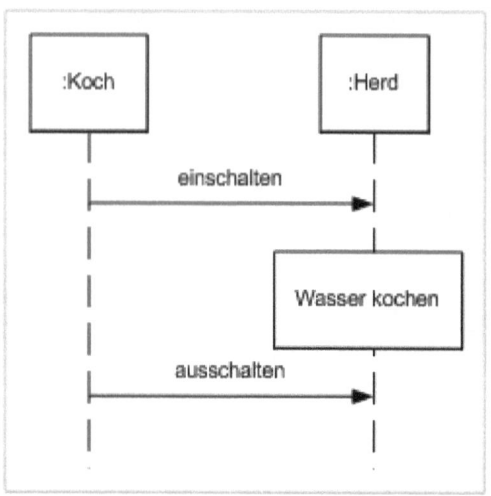

Figure 17-15. Sequence diagram

The Protocol State Machine is a sub-variant of the State Machine. It may be used to model network communication protocols.

Meta modeling

The Object Management Group (OMG) has developed a meta-modeling architecture to define the Unified Modeling Language (UML), called the Meta-Object Facility (MOF). The Meta-Object Facility is a standard for model-driven engineering, designed as a four-layered architecture, as shown in the image at right. It provides a meta-meta model at the top layer, called the M3 layer. This M3-model is the language used by Meta-Object Facility to build meta-models, called M2-models. The most prominent example of a Layer 2 Meta-Object Facility model is the UML metamodel, the model that describes the UML itself. These M2-models describe elements of the M1-layer, and thus M1-models. These would be, for example, models written in UML. The last layer is the M0-layer or data layer. It is used to describe runtime instance of the system.

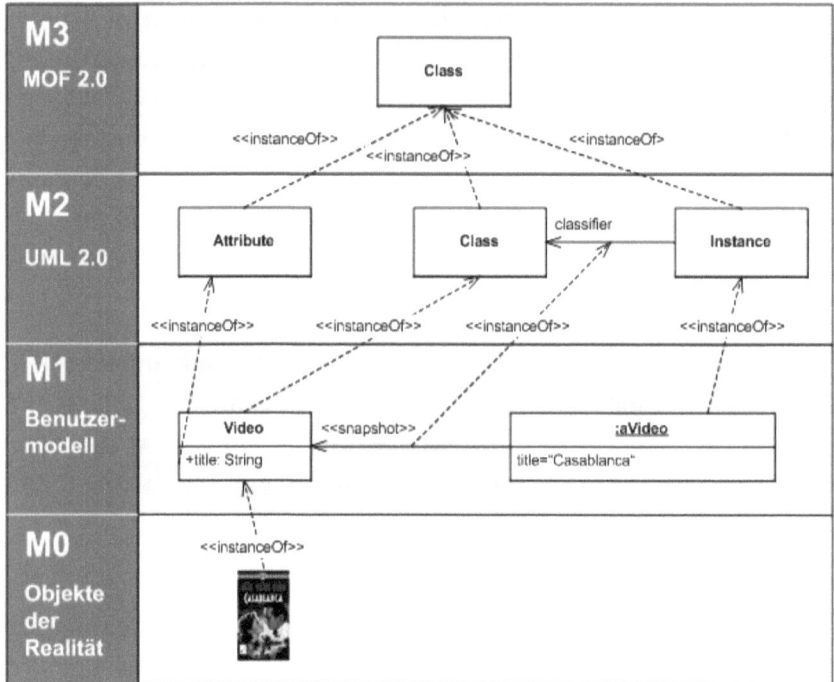

Figure 17-16. Illustration of the Meta-Object Facility.

Beyond the M3-model, the Meta-Object Facility describes the means to create and manipulate models and meta-models by defining CORBA interfaces that describe those operations. Because of the similarities between the Meta-Object Facility M0-model and UML structure models, Meta-Object Facility meta-models are usually modeled as UML class diagrams. A supporting standard of the Meta-Object Facility is XMI, which defines an XML-based exchange format for models on the M3-, M2-, or M1-Layer.

Criticisms

Although UML is a widely recognized and used modeling standard, it is frequently criticized for the following:

Standards bloat
Bertrand Meyer, in a satirical essay framed as a student's request for a grade change, apparently criticized UML as of 1997 for being

unrelated to object-oriented software development; a disclaimer was added later pointing out that his company nevertheless supports UML. Ivar Jacobson, a co-architect of UML, said that objections to UML 2.0's size were valid enough to consider the application of intelligent agents to the problem [178]. It contains many diagrams and constructs that are redundant or infrequently used.

Problems in learning and adopting
The problems cited in this section make learning and adopting UML problematic, especially when required of engineers lacking the prerequisite skills [179]. In practice, people often draw diagrams with the symbols provided by their CASE tool, but without the meanings those symbols are intended to provide.

Linguistic incoherence
The standards have been widely cited as being confused, ambiguous and inconsistent [180] [181]. Even at this late stage, the standard still has many issues [182] [183].

Capabilities of UML and implementation language mismatch
As with any notational system, UML is able to represent some systems more concisely or efficiently than others are. Thus, a developer gravitates toward solutions that reside at the intersection of the capabilities of UML and the implementation language. This problem is particularly pronounced if the implementation language does not adhere to orthodox object-oriented doctrine, as the intersection set between UML and implementation language may be that much smaller.

Dysfunctional interchange format
While the XMI (XML Metadata Interchange) standard is designed to facilitate the interchange of UML models, it has been largely ineffective in the practical interchange of UML 2.x models. This interoperability ineffectiveness is attributable to two reasons. Firstly, XMI 2.x is large and complex in its own right, since it purports to address a technical problem more ambitious than

exchanging UML 2.x models. In particular, it attempts to provide a mechanism for facilitating the exchange of any arbitrary modeling language defined by the OMG's Meta-Object Facility (MOF). Secondly, the UML 2.x Diagram Interchange specification lacks sufficient detail to facilitate reliable interchange of UML 2.x notations between modeling tools. Since UML is a visual modeling language, this shortcoming is substantial for modelers who do not want to redraw their diagrams [184].

Inconsistency

Although a binary relationship line-segment in a class diagram may have a linked Association Class, which allows for attributes to be added and relationships to other classes, an association diamond symbol, necessary for n-ary associations of order > 2 (the most likely candidates for additional attributes and relationships) cannot. The specification also says about class diagrams (mixing two paradigms), that generalizations utilize "intentional" semantics (as a semantic model would), but that other associations use "extensional" semantics, thereby revealing that it is not intended for semantic models but rather for physical record models. Nevertheless, there is no concept in the language for class unique identifier, a concept necessary in such a model.

Illogic

Class models are specified to use "look-across" cardinalities, even though several authors (Merise [185], Elmasri & Navathe [87] amongst others [88]) prefer same-side for roles and both minimum and maximum cardinalities. Recent researchers (Feinerer [89], Dullea et. alia [90]) have shown that this is more coherent when applied to n-ary relationships of order > 2.

In Dullea et.al., "An analysis of structural validity in entity-relationship modeling" one reads "A 'look across' notation such as used in the UML does not effectively represent the semantics of participation constraints imposed on relationships where the degree is higher than binary."

Ingo Feinerer says, "*Problems arise if we operate under the look-across semantics as used for UML associations. Hartmann [91] investigates this situation and shows how and why different transformations fail.*" (Although the "reduction" mentioned is spurious as the two diagrams 3.4 and 3.5 are in fact the same) and also "As we will see on the next few pages, the look-across interpretation introduces several difficulties which prevent the extension of simple mechanisms from binary to n-ary associations."

Inutility

Both use cases and their associated diagrams, especially in projects where these are measured by kilograms or sometimes feet, are often more commonly referred to as "useless cases". Simple user narratives e.g., "what I do at work ..." have shown to be much simpler to record and more immediately useful [186].

Exclusive

The term "Unified" applies only to the unification of the many prior existing and competing Object Orientated languages. Important well known and popular techniques, almost universally used in industry, such as Data Flow Diagrams and Structure charts were not included in the specification.

Modeling experts have written sharp criticisms of UML, including Bertrand Meyer's "UML: The Positive Spin" [187], and Brian Henderson-Sellers and Cesar Gonzalez-Perez in "Uses and Abuses of the Stereotype Mechanism in UML 1.x and 2.0" [188].

UML modeling tools

The most well-known UML modeling tool is IBM Rational Rose. Other tools include, in alphabetical order, ArgoUML, BOUML, Dia, Enterprise Architect, MagicDraw UML, PowerDesigner, Rational Rhapsody, Rational Software Architect, StarUML, and Umbrello. Some of popular development environments also offer UML modeling tools, e.g., Eclipse, NetBeans, and Visual Studio.

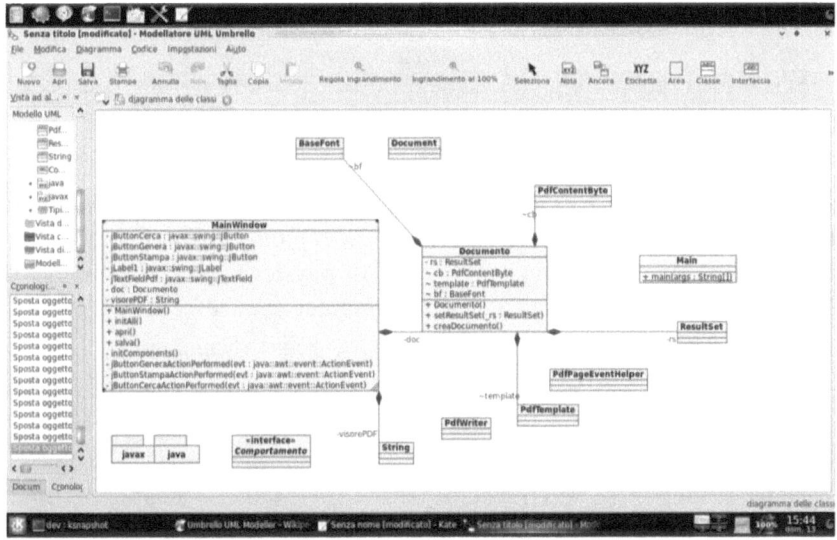

Figure 17-17. Screenshot of Umbrello UML Modeler.

Notes

[1] Extensible programming is a term used in computer science to describe a style of computer programming that focuses on mechanisms to extend the programming language, compiler and runtime environment. Extensible programming languages, supporting this style of programming, were an active area of work in the 1960s, but the movement was marginalized in the 1970s.

[2] A profile in the Unified Modeling Language (UML) provides a generic extension mechanism for customizing UML models for particular domains and platforms. Extension mechanisms allow refining standard semantics in strictly additive manner, so that they cannot contradict standard semantics.

[3] A stereotype is one of three types of extensibility mechanisms in the Unified Modeling Language (UML). They allow designers to extend the vocabulary of UML in order to create new model elements, derived from existing ones, but that have specific properties that are suitable for a particular problem domain or otherwise specialized usage. The nomenclature is derived from the original meaning of stereotype, used in printing. For example, when modeling a network you might need to have

symbols for representing routers and hubs. By using stereotyped nodes you can make these things appear as primitive building blocks.

[4] James E. Rumbaugh (born 24 September 1947) is an American computer scientist and object methodologist who is best known for his work in creating the Object Modeling Technique (OMT) and the Unified Modeling Language (UML). Rumbaugh has a B.S. in physics from MIT, an M.S. in astronomy from Caltech, and a Ph.D. in computer science from MIT.

[5] Object-oriented analysis and design (OOAD) is a software engineering approach that models a system as a group of interacting objects. Each object represents some entity of interest in the system being modeled, and is characterized by its class, its state (data elements), and its behavior. Various models can be created to show the static structure, dynamic behavior, and run-time deployment of these collaborating objects. There are a number of different notations for representing these models, such as the Unified Modeling Language (UML).

[6] Grady Booch (born February 27, 1955) is an American software engineer, and Chief Scientist, Software Engineering in IBM Research. Booch is best known for developing the Unified Modeling Language with Ivar Jacobson and James Rumbaugh. He earned his bachelor's degree in 1977 from the United States Air Force Academy and a master's degree in electrical engineering in 1979 from the University of California, Santa Barbara.

[7] Ivar Hjalmar Jacobson (born 1939) is a Swedish computer scientist, known as major contributor to UML, Objectory, Rational Unified Process (RUP) and aspect-oriented software development.

[8] Objectory AB, known as Objectory System, was founded in 1987 by Ivar Jacobson. In 1991, it was acquired and became a subsidiary of Ericsson.

[9] OOPSLA (Object-Oriented Programming, Systems, Languages & Applications) is an annual ACM research conference. OOPSLA mainly takes place in the United States, while the sister conference of OOPSLA, ECOOP, is typically held in Europe. It is operated by the Special Interest Group for Programming Languages (SIGPLAN) group of the Association for Computing Machinery (ACM).

Chapter 18. DoDAF

DoDAF is the Department of Defense Architecture Framework. One can view DoDAF products, or at least the 3 views, as ANSI/IEEE 1471-2000 or ISO/IEC 42010 viewpoints. DoDAF consists of a set of Artifact Views [57].

DoDAF views are organized into four basic view sets:

- overarching All View (AV),
- Operational View (OV),
- Systems View (SV),
- Technical Standards View (TV).

Only a subset of the full DoDAF viewset is usually created for each system development.

DoDAF overview

DoDAF is the implementation chosen by the United States Department of Defense to gain compliance with the Clinger-Cohen Act and United States Office of Management and Budget Circulars A-11 and A-130. It is administered by the Undersecretary of Defense for Business Transformation's DoDAF Working Group. DoDAF was formerly named C4ISR (Command, Control, Communications, Computers, Intelligence, Surveillance and Reconnaissance) AF. Other derivative frameworks based on DoDAF include the NATO Architecture Framework (NAF) and Ministry of Defence (United Kingdom) Architecture Framework (MODAF).

Representation

Representations for the DoDAF products may be drawn from many diagramming techniques including: tables, ICAM Definition Language, Entity-Relationship Diagrams (ERDs), UML/ SysML, and other custom techniques depending on the product, tool used, and contractor/customer preferences. There is a UPDM (UML Profile for DoDAF and MODAF) effort within the OMG to standardize the representation of DoDAF products when UML is used (Released August 2007).

DoDAF generically describes the representation of the artifacts to be generated, but allows considerable flexibility regarding the specific formats and modeling techniques. The DoDAF deskbook provides examples in using traditional systems engineering and data engineering techniques, and secondly, UML format. DoDAF proclaims latitude in work product format, without professing one diagramming technique over another.

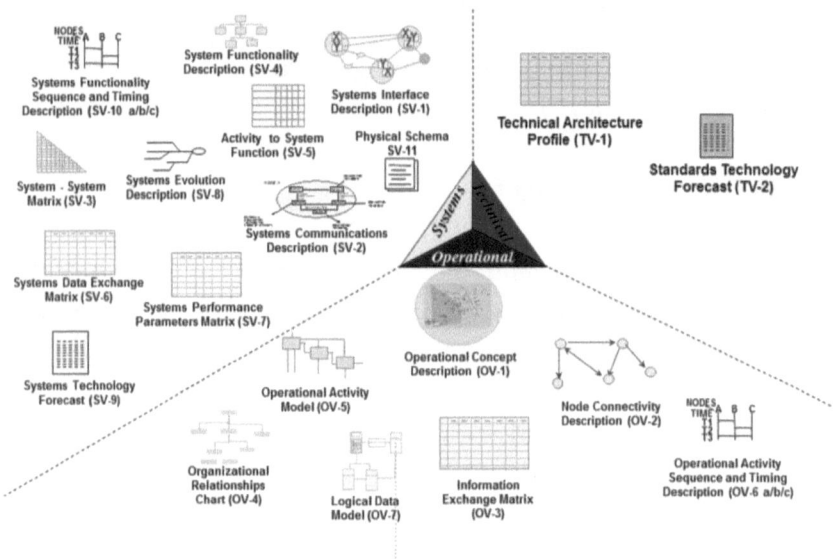

Figure 18-1. DoDAF Overview

Table 18-1. Mapping of UML to Architecture Framework

View	Description	Derived	Displayed in
OV-1	Operational Concept Graphic	Not Derived from any UML Diagram must be developed separately.	Class Diagram with Icons
OV-2	Operational	Object Diagram showing the	Class

View	Description	Derived	Displayed in
	Node Connectivity	information diagram between nodes. Based on a Class Diagram with association classes.	Diagram or Collaboration Diagram
OV-3	Information Exchange Matrix	Derived from OV-2s and Activity Diagrams showing the information exchange between operational activities.	Excel
OV-4	Org Chart	Not normally developed in UML modeling, but the Class Diagram can be used to display the information.	Class Diagram
OV-5	Activity Model	Use the UML Activity Diagram with information exchanges on the arcs.	Activity Diagram for operational nodes
OV-6a	Operational Rules Model	Can be autogenerated from a UML State Diagram	
OV-6b	State Transition	Use UML State Transition Diagram Directly	State Transition
OV-6c	Operational Events	Use UML Sequence Diagram Directly	Sequence Diagram
OV-7	Logical Data Model	Use the UML Class Diagram for Logical Data Modeling Directly	Class Diagram
SV-1	System Interface	Use UML Class and Object Diagrams Can also use the Deployment Diagram for the high level.	Class Diagram
SV-2	Communications	Same as SV-1 but at lower level of detail	Class Diagram
SV-3	System-to-System Interface	Autogenerated from the SV-1 and SV-2	Matrix
SV-4	System Functionality Description	Use Activity Diagrams same as OV-5 but with system rather than operational functions.	Activity
SV-5	System Functions to Operational	Use Activity Diagrams same as OV-5 but with system rather than operational functions.	Activity

View	Description	Derived	Displayed in
	Activities		
SV-6	System Data Exchange	Same as OV-3 but with system rather than operational functions.	Matrix
SV-7	System Performance Parameters	Knowledge outside of UML diagrams, [but can be captured in tagged value sets and the matrix generated from a script]	Matrix
SV-8	System Evolution	Knowledge outside of UML modeling, but the UML Class Diagram can be used to display the information.	Class Diagram
SV-9	System Technology Forecast	Outside of UML Domain [Class diagrams and tagged values can be used]	Word
SV-10		Same as OV-6 but at system rather than operational level	State & Sequence diagrams
SV-11	Physical Data Model	Same as OV-7 but at the physical data model level.	Class diagram
TV-1	Technical Architecture Profile	Outside of UML Domain [Class diagrams and tagged values can be used]	Word
TV-2	Technical Standards Forecast	Exclusive of UML modeling	Word
AV-1	Overview and Summary Information	Exclusive of UML modeling	Word
AV-2	Integrated Dictionary	Generated from the UML model, all diagrams.	TBD

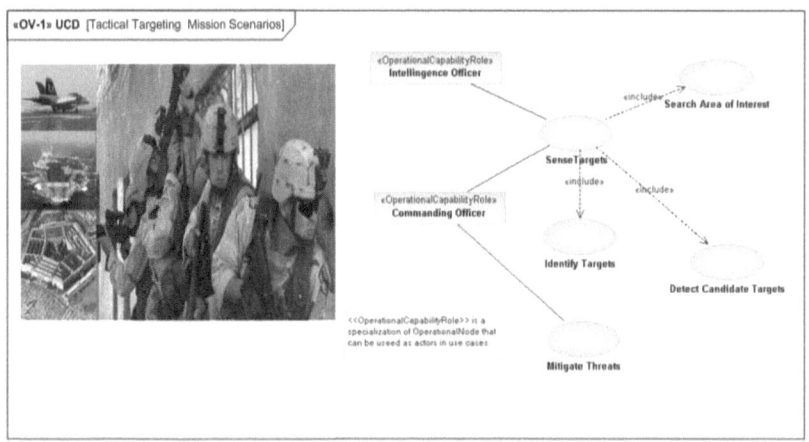

Figure 18-2. OV-1 Captures the Operational Context graphically and as Role and Use Case models representing Mission Context, Scenarios and enabling Capabilities

Figure 18-3. OV-5 Activity Diagram captures the behavior defined by OV-1 Use Cases via Operational Activities subsequently allocated to Operational Nodes

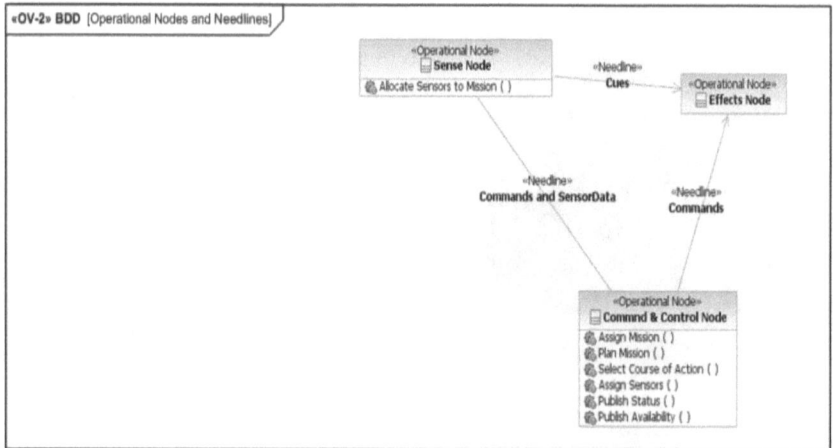

Figure 18-4. OV-2 Operational Nodes and Needlines as aggregations of allocated Operational Activities and Information Flows respectively

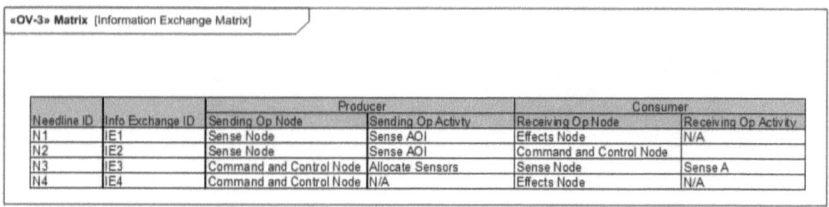

Figure 19-5. OV-3 Information Exchange Matrix auto generated based on Operational Activity Information Flows

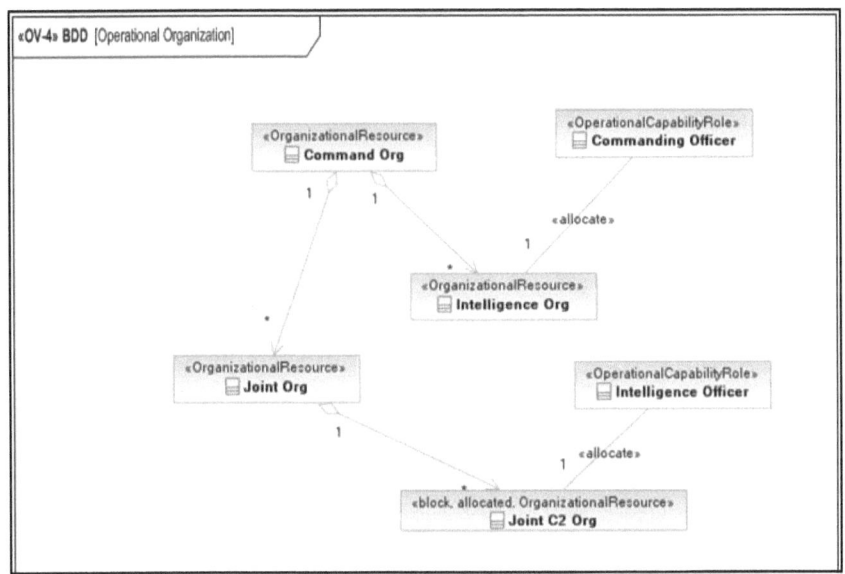

Figure 18-6. OV-4 Organizations and Roles allocated to each Organization are modeled by UML Composite Structure and SysML and Block Definition Diagrams

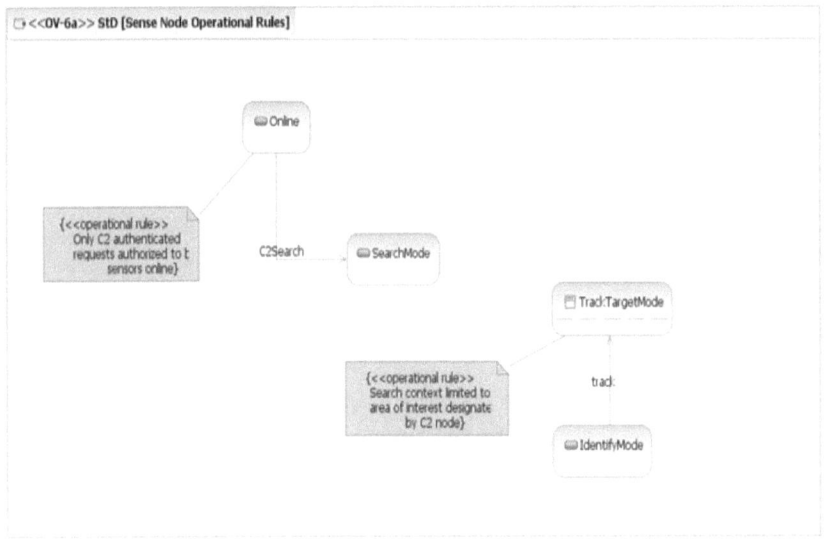

Figure 18-7. OV-6a Captures the Operational Rules allocated to Operational model elements including Nodes, Activities, Flows, Information Elements and States

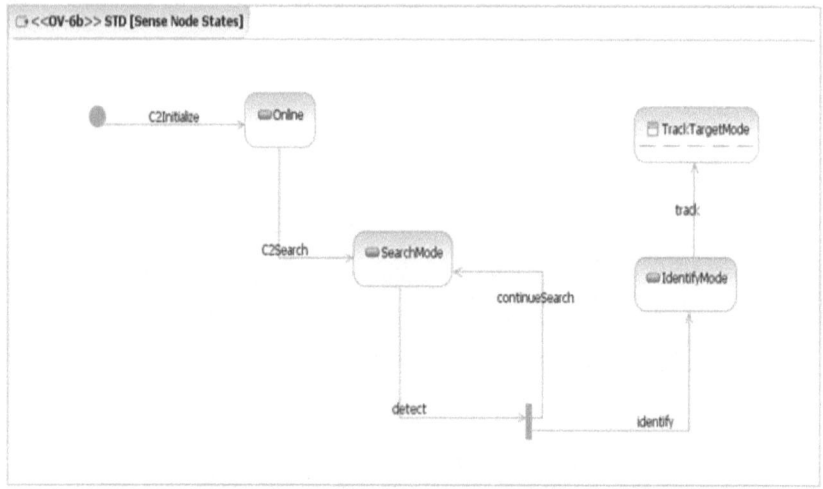

Figure 18-8. OV-6b State Trace Diagram captures the Operational States and Transitions within Operational Nodes

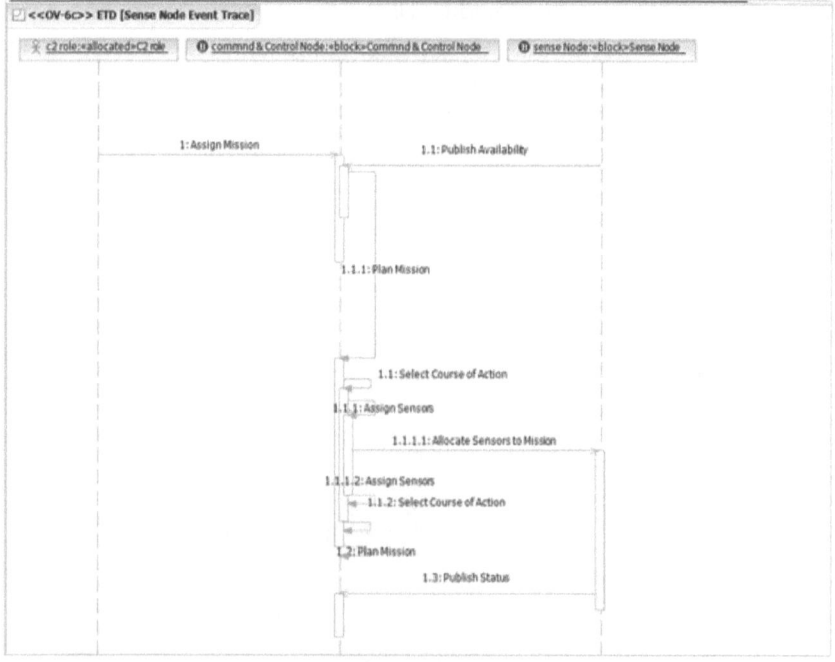

Figure 18-9. OV-6c Event Trace diagram captures the event flow (messages) between instances of Operational Nodes...synchronized with Operational Activities allocated to Nodes

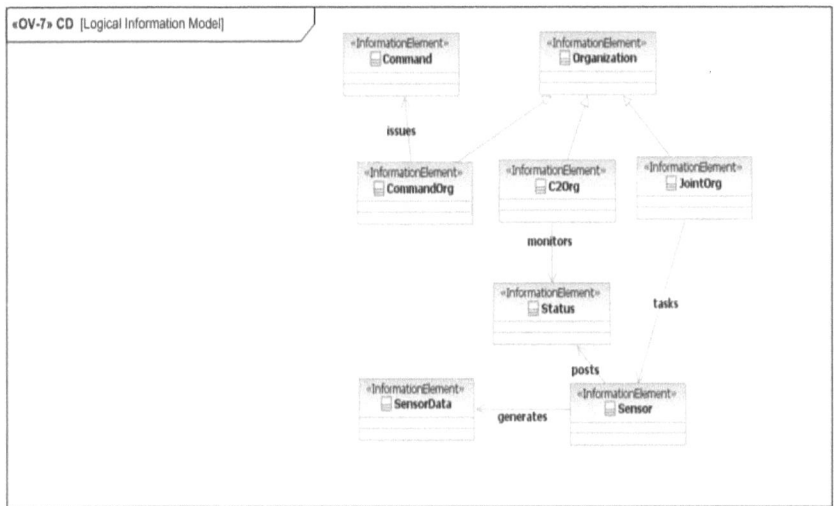

Figure 18-10. OV-7 Defines Information Model for all Operational Information Flow elements identified within the integrated model and viewed via OV-3 and OV-5

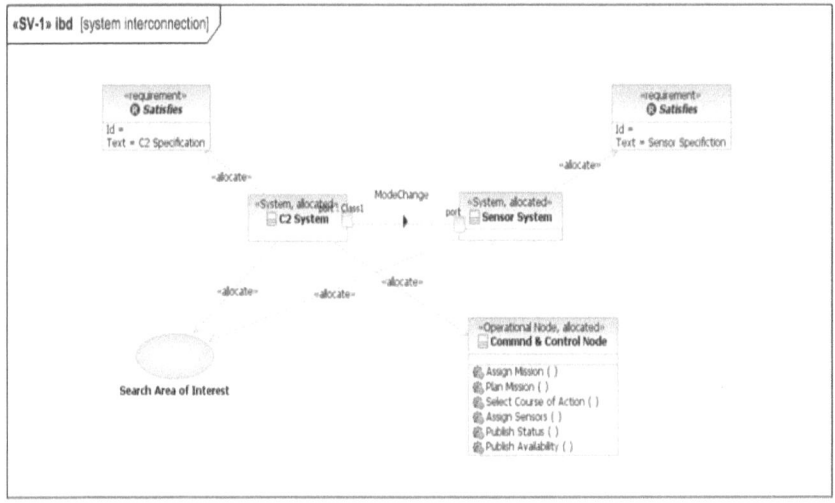

Figure 18-11. SV-1 Provides the view of the Systems, their connections and data flows, and allocations to Mission Scenarios and Operational Activities

Chapter 19. Technique Evaluation and Selection

Because the conceptual modeling method can sometimes be purposefully vague to account for a broad area of use, the actual application of concept modeling can become difficult. To alleviate this issue, and shed some light on what to consider when selecting an appropriate conceptual modeling technique, the framework proposed by Gemino and Wand [189] will be discussed in the following text. However, before evaluating the effectiveness of a conceptual modeling technique for a particular application, an important concept must be understood: Comparing conceptual models by way of specifically focusing on their graphical or top level representations is shortsighted [189]. Gemino and Wand make a good point when arguing that the emphasis should be placed on a conceptual modeling language when choosing an appropriate technique. In general, a conceptual model is developed using some form of conceptual modeling technique. That technique will utilize a conceptual modeling language that determines the rules for how the model is arrived at. Understanding the capabilities of the specific language used is inherent to properly evaluating a conceptual modeling technique, as the language reflects the techniques descriptive ability. In addition, the conceptual modeling language will directly influence the depth at which the system is capable of being represented, whether it is complex or simple [189].

Considering Affecting Factors

Building on some of their earlier work [190], Gemino and Wand acknowledge some main points to consider when studying the affecting factors: the content that the conceptual model must represent, the method in which the model will be presented, the characteristics of the model's users, and the conceptual model languages specific task [189]. The conceptual model's content should be considered in order to select a technique that would allow relevant information to be presented. The presentation method for selection purposes would focus on the techniques ability to represent the model at the intended level of depth and detail. The characteristics of the model's users or participants are an important aspect to consider. A participant's background and experience should coincide with the conceptual models complexity; else, misrepresentation of the system or misunderstanding of key system concepts could lead to problems in that systems realization. The conceptual model language task will further allow an appropriate technique to be chosen. The difference between creating a system conceptual model to convey system functionality and creating a system conceptual model to interpret that functionality could involve to completely different types of conceptual modeling languages.

Considering Affected Variables

Gemino and Wand go on to expand the affected variable content of their proposed framework by considering the focus of observation and the criterion for comparison [189]. The focus of observation considers whether the conceptual modeling technique will create a "new product", or whether the technique will only bring about a more intimate understanding of the system being modeled. The criterion for comparison would weigh the ability of the conceptual modeling technique to be efficient or effective. A conceptual modeling technique that allows for development of a system model which considers all system variables at a high level may make the process of understanding the system functionality more efficient, but the technique lacks the necessary information to explain the internal processes, rendering the model less effective.

When deciding which conceptual technique to use, the recommendations of Gemino and Wand can be applied in order to evaluate properly the scope of the conceptual model in question. Understanding the conceptual models scope will lead to a more informed selection of a technique that properly addresses that particular model. In summary, when deciding between modeling techniques, answering the following questions would allow one to address some important conceptual modeling considerations.

1. What content will the conceptual model represent?
2. How will the conceptual model be presented?
3. Who will be using or participating in the conceptual model?
4. How will the conceptual model describe the system?
5. What is the conceptual models focus of observation?
6. Will the conceptual model be efficient or effective in describing the system?

Evaluation Criteria for a Simulation Conceptual Model

There are four primary evaluation criteria for a simulation conceptual model:

1. Completeness: the simulation conceptual model identifies all representational entities and processes of the problem domain, the "mission space" in DoD parlance, and all control and operating characteristics of the simulation, "simulation space," needed to ensure that specifications for the simulation fully satisfy simulation requirements.
2. Consistency: representational entities and processes within the conceptual model are addressed from compatible perspectives in regard to such features as coordinate systems and units, levels of aggregation/disaggregation (which is of particular importance in C4ISR simulation), precision, accuracy, and descriptive paradigms.
3. Coherence: the conceptual model is organized so that all elements of both mission space and simulation space have function (i.e., there are not extraneous items) and potential (i.e., there are no parts of the conceptual model that are impossible to activate).

4. Correctness: the simulation conceptual model is appropriate for the intended application and has potential to perform in such a way as to satisfy fully simulation requirements.

A Multi-Perspective Framework for Evaluation

Our brief overview of the state of the art in evaluating conceptual models reveals a number of peculiarities. First, there has been only little work on the explicit evaluation of reference models[1]. The majority of related work is concentrated on evaluating conceptual models or—to a lower extent—on evaluating modeling languages. Most authors suggest a multi-perspective approach. Perspectives are often inspired by linguistic categories (syntax, semantics and pragmatics), sometimes extended by a more differentiated consideration of users' perception or a model's relationship to reality. While most frameworks include the judgment of language features, some lack an explicit differentiation of meta- and object levels. The use of ontologies is a valuable contribution to a more comprehensive and obliging evaluation. However, usually the ontology that serves as a reference is taken for granted, thereby terminating the course of reasoning in a somewhat ideological way.

An objective evaluation is hard to accomplish. Hence, the idea is to get closer to objectivity by fostering a more differentiated and balanced judgment. In this sense, the structure that is suggested here is an attempt, not the solution—or, following Wittgenstein [191]—a structure, not the structure. The conceptual framework includes four main perspectives, which are structured in a number of specific aspects. The perspectives are not necessarily independent. Their differentiation is mainly motivated by analytical reasons. The economic perspective is aimed at discussing criteria that are relevant for judging costs and benefits related to the use of conceptual models. Among other things, it takes into account protection of investment, possible effects on information quality and competitiveness. The deployment perspective is focused on criteria that are relevant for those who work with the models. It stresses criteria such as comprehensibility, compatibility with other representations being used in an organization, availability of tools, etc. Conceptual models are artifacts that have been designed for a certain purpose. In addition, they will usually be related to the analysis and design of information systems. The engineering perspective is aimed at evaluating a conceptual model as a design artifact that has to satisfy a specification—including the support for analysis and transformation. With respect to their claim for general validity, reference models resemble scientific theories. The epistemological perspective is aimed at evaluating reference models as the results of scientific research.

The evaluation of a conceptual model depends also on its type—e.g., an object model, a data model, a business process model, etc. However, due to the limited space of this chapter, specific features of particular model types will be widely abstracted in form. The suggested criteria are intended to provide guidance for evaluating conceptual models. They do not im ply a specific scale level. Most of them will allow for classification, some for applying an ordinal scale only, e.g., a Likert scale[2]. If there is need for calculating aggregated evaluation measures, one could define corresponding higher order scales. However, this would cause a distortion of the evaluation result.

These suggestions are based on previous work on the evaluation of modeling languages. In Frank and Prasse [192], a framework for the evaluation of object-oriented modeling languages is presented. It includes 33 criteria, which are applied to a comparison of the UML and the OML [193]. Frank [194] suggests a multi-perspective framework for the discursive evaluation of modeling languages. Frank and Lange [195] are aimed at languages for modeling business processes. It presents a comprehensive analysis of requirements for these kinds of languages.

The Economic Perspective

Both, the construction and the (re-) use of conceptual models chiefly depend on economic aspects. We will mainly take the viewpoint of a potential model user rather than that of a model developer. The type of user depends on the purpose for which a conceptual model is deployed. Some will take a conceptual model as a foundation for developing software (pre-development use—referred to as type 1 in the table that illustrates the framework). For other users, a conceptual model serves mainly as a documentation of existing software (post-development use—referred to as type 2). Both pre- and post-development use can be applied to object models (or data models respectively) or business process models. In the case of post-development use, component models or application models—which would mainly focus on interfaces—are an option as well? In order to illustrate their deployment, they should be integrated with business process models. A third group of potential users is primarily interested in organizational or strategic issues (business (re-) design-referred to as type 3). Conceptual models that represent corporate strategies or organizations (e.g., business processes and organizational charts) are suitable for this category of use. In a particular case, different approaches to using a conceptual model may be combined, for instance pre-development use and business (re-) design.

The Engineering Perspective

A conceptual model is a design artifact that can be regarded as a specification of possible solutions to a range of problems. From an engineering viewpoint, two questions are pivotal: Does the model fulfill the requirements to be taken into account? Is the specification suited for supporting the intended purposes of the model? To analyze these questions, four aspects are differentiated: definition, explanation, language features, model features.

Testing a model against requirements implies the requirements are to be made explicit in a comprehensive and precise way. Requirements include a definition of the intended application domains as well as a definition of the purposes to be satisfied. In the ideal case, these definitions should allow for deciding whether the model fits a particular application area or whether it supports a certain purpose. Note, however, that this does not only depend on the quality of the requirements documentation. Furthermore, every prospective user should know the requirements and purposes of the application he has in mind. In addition to merely defining the requirements, the model should also be explained in the sense that a potential user is supported in understanding and judging it. This includes an assignment of model elements to requirements as well as a substantiation of major design decisions that the model is based on. Often, design decisions require a compromise. This should be discussed including the resulting drawbacks. With respect to a modeling language, the following criteria are relevant: level of formalization, extensibility, supported conceptual views, integration of views, tool support and concepts to support the adaptation of models. Technical features of a model include formal correctness, model architecture and adaptability.

The Deployment Perspective

The success of a reference model depends heavily on its users. This includes their ability as well as their willingness to deal with the model. Within this perspective, the framework includes the following aspects: understandability, appropriateness and attitude. In order to foster communication between the involved stakeholders, a model should be understandable. In other words, it should correspond to concepts, the prospective model with which the users are familiar. A conceptual model should stress an appropriate level of abstraction in detail—with respect to the purpose, a model is supposed to fulfill. If prospective users are not willing to make use of the model or if there are any objections against the model's usability, this lack of attitude can become a critical success factor. Therefore, it should be taken into account, even if it does not necessarily correspond directly to certain model features.

The Epistemological Perspective

This perspective serves to enrich the evaluation of conceptual models with epistemological considerations. They are differentiated into four interrelated aspects: the evaluation of theories, general principles of scientific research, critical reflection of human judgment and reconstruction of scientific progress.

Conceptual models reveal similarities to scientific theories. Like theories, they are supposed to provide representations not just of a single instance (an enterprise, an application, etc.), but of an entire class. In addition, they can be regarded as contributions to the body of knowledge within a certain domain of interest. Therefore, it makes sense to apply criteria that are used for the evaluation of theories to the evaluation of conceptual models. There is, however, one major difference between theories and conceptual models. A theory is aimed at describing the world as it is. Hence, a key criterion for assessing theories is truth, or rather: a certain concept of truth, such as the correspondence, coherence or consensus concept. A concept of truth is only of limited use for evaluating conceptual models, since they are usually aimed at intended systems or future worlds: They are not only descriptive, but also prescriptive. Nevertheless, also with conceptual models, the claim for truth cannot be entirely neglected—we could speak of "relaxed truth": A conceptual model does not have to fit reality entirely; however, it should not contradict evidence. Hence, the descriptive parts of the model and the assumptions underlying the prescriptive parts can be evaluated according to the judgment of theories.

The correspondence concept of truth recommends testing a hypothesis against reality. This requires a precise description of the model and its intended applications as well as testing procedures that allow for comparing a statement with perceptions of reality. The coherence concept of truth recommends that a new hypothesis should be in line with an established body of knowledge, e.g., with research results and opinions found in acknowledged publications.

Applied to reference models, this implies that assumptions underlying the design of a model should not contradict accepted knowledge, e.g., established accounting principles (notice that this is just one notion of truth). From the viewpoint of truth as result of a consensus, emphasis is on discursive judgment by experts. This would recommend getting acknowledged people involved who should discuss and eventually confirm the assumptions a reference model is based on.

Despite the ongoing discussion on concepts of truth and corresponding research methods, there are three generic principles that allow for differentiating scientific research from other sources of knowledge: abstraction, originality and judgment. While not necessarily with the same rigor, they should apply to reference models, too. A high quality reference model should abstract from peculiarities of single instances and from changes that may occur over time. Abstraction, however, does not simply mean to fade arbitrarily out parts of the domain.

Instead, abstraction should be made explicit and should include hints of how to turn it into a concrete description that applies to a particular case. With conceptual models, originality is hard to judge. Nevertheless, it is certainly important. This is especially the case for conceptual models that result from scientific research (see "progress"). Judgment in science means that there has to be given comprehensive reason/justification for any hypothesis. For this purpose, one will usually refer to the preferred concept of truth and the related testing procedures. This can be applied to the descriptive parts of a conceptual model, too. With respect to decisions that motivate prescriptive elements of a conceptual model, this is different, because truth is not the issue. In order to provide reasons for design decisions, reference could be made to the accepted state of the art (following the coherence concept) or to discursive judgment by experts (following the consensus concept). In any case, judgment implies that every non-trivial assumption that design decisions are based on should be made explicit and reasons given for the choice.

Epistemology deals with the study of scientific judgments or, in other words, with the limits of human knowledge. Despite the ongoing debate, a critical or even skeptical evaluation of our perception and ability to judge prevails. There are many kinds of deception. With respect to the social sciences (or the humanities), perception and judgment are often biased by social/cultural constructions one is not entirely aware of. With respect to conceptual models, there is even more reason for epistemological skepticism. Conceptual models are linguistic artifacts: They are described using a language and—on another level of abstraction—they represent a language themselves. Although we are able to reflect upon language, for instance by distinguishing between object and meta-level language, our ability to speak and understand a language is commonly regarded as a competence that we cannot entirely comprehend. Therefore, any research that aims at inventing new "language games" (i.e., artificial languages and actions built upon them), has to face a subtle challenge: Every researcher is trapped in a network of language, patterns of thought and action he or she cannot completely transcend, leading to a paradox that can hardly be resolved.

Understanding a language is not possible without using a language. At the same time, any language we use for this purpose will bias our perception and judgment or, as the early Wittgenstein put it, "The limit of my language means the limit of my world" [191]. If one has to judge a reference model specified in UML and happens to dislike UML, an objective evaluation of the model will be hardly possible. In addition, if a conceptual model of an accounting system makes use of terms that are different from those we use for accounting, it is very likely that we do not find it comprehensive—although it might be superior with respect to consistency or adaptability.

While it seems impossible to overcome entirely these obstacles, they can be met with certain precautions. Everyone involved in the evaluation of a conceptual model should name the modeling languages he or she is familiar with as well as preferences for modeling languages and technical languages. Then, everyone should reflect upon the question how his or her language background could influence their judgment. This could contribute to a more critical distance and a more objective judgment.

Notes

[1] A Reference model in systems, enterprise, and software engineering is a model of something that embodies the basic goal or idea of something and can then be looked at as a reference for various purposes.

[2] A Likert scale is a psychometric scale commonly involved in research that employs questionnaires. It is the most widely used approach to scaling responses in survey research, such that the term is often used interchangeably with rating scale, or more accurately the Likert-type scale, even though the two are not synonymous. The scale is named after its inventor, psychologist Rensis Likert.

Works Cited

[1] D. K. Pace, "Ideas About Simulation Conceptual Model Development," *John Hopkins APL Technical Digest,* vol. 21, no. 3, pp. 327-336, 2000.

[2] D. Isbell, M. Hardin and J. Underwood, "Mars Climate Orbiter Team Finds Likely Cause of Loss," Mars Polar Lander, 30 09 1999. [Online]. Available: http://mars.jpl.nasa.gov/msp98/news/mco990930.html. [Accessed 10 10 2011].

[3] C. K. A. Solvberg, "Activity Modeling and Behavior Modeling," in *Proceedings of the IFIP WG 8.1 working conference on comparative review of information systems design methodologies: improving the practice*, Amsterdam, 1986.

[4] J. A. Sokolowski and C. M. Banks, Modeling and Simulation Fundamentals: Theoretical Underpinnings and Practical Domains, 1 ed., Wiley, 2010.

[5] H. J. Miser and E. S. Quade, Handbook of Systems Analysis: Cases, Vol. 3, John Wiley & Sons, 1996.

[6] R. O. Lewis and G. Q. Coe, "A Comparison Between the CMMS and the Conceptual Model of the Federation," in *97 Fall Simulation Interoperability Workshop Papers*, 1997.

[7] J. Sheehan, T. Prosser, H. Conley, G. Stone, K. Yentz and J. Morrow, "Conceptual Models of the Mission Space (CMMS): Basic Concepts, Advanced Techniques, and Pragmatic Examples," in *98 Spring Simulation Interoperability Workshop Papers*, 1998.

[8] T. H. Johnson, "Mission Space Model Development, Reuse and the Conceptual Models of the Mission Space Toolset," in *98 Spring Simulation Interoperability Workshop Papers*, 1998.

[9] P. J. Roache, Verification and Validation in Computational Science and Engineering, Hermosa Press, 1998.

[10] P. P. (. Chen, "Data & Knowledge Engineering, ,," Elsevier Science website, 12 10 2011. [Online]. Available: http://www.elsevier.com/wps/find/journaldescription.cws_home/505608/description#description. [Accessed 14 10 2011].

[11] G. N. Hone and M. R. Moulding, "Developments in Application Domain Modeling for the Verification and Validation of Synthetic Environments: Training Process and System Definition," in *Proceedings of the Spring 1999 Simulation Interoperability Workshop*, Orlando, FL, 1999.

[12]　B. P. Zeigler, T. G. Kim and H. Praehofer, Theory of Modeling and Simulation, 2nd ed., New York: Academic Press, 1999.

[13]　J. L. Casti, Alternative Realities – Mathematical Models of Nature and Man, John Wiley & Sons, 1989.

[14]　I. Davies, P. Green, M. Rosemann, M. Indulska and S. Gallo, "How do practitioners use conceptual modeling in practice?," *Data & Knowledge Engineering*, vol. 58, pp. 358-380, 2006.

[15]　R. Hirschheim, H. K. Klein and K. Lyytinen, Information Systems Development and Data Modeling: Conceptual and Philosophical Foundations, Cambridge: Cambridge University Press, 1995.

[16]　P. B. Checkland, Systems Thinking, Systems Practice, Wiley, 1981.

[17]　E. J. Davidson, "Joint application design (JAD) in practice," *Journal of Systems & Software*, vol. 45, pp. 215-223, 1999.

[18]　T. Halpin, "Object-Role Modeling: an overview," Microsoft Corporation, 2001.

[19]　S. Smith, "SSADM (Structured Systems Analysis & Design Method)," SearchSoftwareQuality.com, 09 1999. [Online]. Available: http://searchsoftwarequality.techtarget.com/definition/SSADM. [Accessed 14 10 2011].

[20]　OGC, "The National Archives," 04 03 2009. [Online]. Available: http://webarchive.nationalarchives.gov.uk/20100503135839/http:// www.ogc.gov.uk/index.asp. [Accessed 11 06 2011].

[21]　M. Goodland and K. Riha, "History of SSADM," 20 01 1999. [Online]. Available: http://www.dcs.bbk.ac.uk/~steve/1/sld005.htm. [Accessed 06 11 2011].

[22]　Model Systems, "Model Systems and SSADM," 08 03 2002. [Online]. Available: http://web.archive.org/web/20090402163313/http://www.modelsys.com/msssadm.htm. [Accessed 06 11 2011].

[23]　SSADM foundation, Business Systems Development with SSADM, The Stationery Office, 2000.

[24]　P. B. Checkland and J. Poulter, Learning for Action: A short definitive account of Soft Systems Methodology and its use for Practitioners, teachers and Students, Chichester: Wiley, 2006.

[25]　P. B. Checkland, Systems Thinking, Systems Practice, John Wiley & Sons Ltd, 1998.

[26]　B. Wilson, Systems: Concepts, Methodologies and Applications, John Wiley & Sons Ltd., 1990.

[27]   P. B. Checkland and J. Scholes, Soft Systems Methodology in Action, John Wiley & Sons Ltd., 1990.

[28]   P. B. Checkland, "Soft Systems Methodology: A thirty year retrospective," *Systems Research and Behavioral Science,* vol. 17, p. 11–58, 2000.

[29]   P. B. Checkland, "Soft Systems Methodology," in *Rational Analysis for a Problematic World Revisited,* Chichester, Wiley, 2001.

[30]   G. Seymour, Harry's Game, HarperCollins Publishers Ltd, 1975.

[31]   F. H. Gregory, "Cause, Effect, Efficiency & Soft Systems Models," *Journal of the Operational Research Society,* vol. 44, no. 4, pp. 149-168, 1993.

[32]   C. Jarvis, "Business Open Learning Archive," BOLA Project, 09 04 2009. [Online]. Available: http://www.bola.biz/research/ssm.html. [Accessed 2011 7 10].

[33]   S. K. Probert, "Logic and conceptual modeling in Soft Systems Methodology," in *Proceedings of the Conference on the Theory, Use and Integrative Aspects of IS Methodologies,* 1993.

[34]   P. B. Checkland and S. Holwell, Information, Systems and Information Systems, Chichester: John Wiley & Sons, 1998.

[35]   G. Curtis, Business Information Systems. Second Edition, Addison-Wesley Ltd., 1995.

[36]   J. Mingers and S. Taylor, "The use of Soft Systems Methodology in Practice," *Journal of the Operational Research Society,* vol. 43, no. 4, 1992.

[37]   B. Wilson and K. v. Haperen, "Improving regional policing: a review of protective services," *International Journal of Police Science & Management,* vol. 12, no. 2, 2010.

[38]   D. Avison and A. T. Wood-Harper, Multiview: An Exploration in Information Systems Development, Blackwell Scientific Publications, 1990.

[39]   F. H. Gregory, "Logic and Meaning in Conceptual Models: Implications for Information System Design," *Systemist,* vol. 15, no. 1, 1992.

[40]   F. H. Gregory, "SSM for Knowledge Elicitation & Representation," *Journal of the Operational Research Society,* vol. 46, pp. 562-578, 1995.

[41]   J. H. Klein, "Cognitive processes and operational research: a human information processing perspective," *Journal of the Operational Research Society,* vol. 45, no. 8, 1994.

[42]   J. H. Klein, "Over-simplistic cognitive science: A response," *Journal of the Operational Research Society*, vol. 46, no. 4, pp. 275-6, 1995.

[43]   L. A. Zadeh, "Fuzzy Sets," *Information and Control*, vol. 8, pp. 338-353, 1965.

[44]   L. A. Zadeh, "Outline of a New Approach to the Analysis of Complex Systems and Decision Processes," *IEEE Transactions on Systems, Man and Cybernetics*, vol. 3, no. 1, pp. 28-44, 1973.

[45]   L. A. Zadeh, "Possibility theory and soft data analysis," in *Mathematical frontiers of the social and policy sciences*, L. Cobb and R. M. Thrall, Eds., Boulder, CO: Westview, 1981, pp. 69-129.

[46]   J. W. Forrester, Industrial dynamics, Cambridge: The MIT Press, 1961.

[47]   T. D. Clark, "The system dynamics approach to analysis of complex industrial and management systems," in *Proceedings of the 15th conference on Winter Simulation - Volume 2*, Piscataway, 1983.

[48]   D. C. Montgomery, Design and Analysis of Experiments, Wiley, 2008.

[49]   J. S. Strickland, Mathematical Models of Warfare and Combat Phenomenon, Lulu.com, 2011.

[50]   G. A. Miller, "The Magical Number Seven, Plus or Minus Two: Some Limits on our Capacity for Processing Information," *Psychological Review*, vol. 63, pp. 81-97, 1956.

[51]   T. Gilbert, "FCS Evaluation criterea for technology assessment," in *Integrated CNS Technologies Conference & Workshop*, NASA Glenn Research Center, 2006.

[52]   E. Yourdon, Managing the Structured Techniques: Strategies for Software Development in the 1990s., Yourdon Press, 1986.

[53]   FAA, "FAA System Safety Handbook, Appendix D," 30 12 2000. [Online]. Available: http://www.faa.gov/library/manuals/aviation/risk_management/ss_handbook/media/app_d_1200.pdf. [Accessed 08 10 2011].

[54]   D. Levitt, "Introduction to Structured Analysis and Design. Retrieved 21 Sep 2008," 21 09 2008. [Online]. Available: http://faculty.inverhills.edu/dlevitt/CS%202000%20(FP)/Introductio n%20to%20Structured%20Analysis%20and%20Design.pdf. [Accessed 08 10 2011].

[55]   G. Salvendy, Handbook of Industrial Engineering: Technology and Operations Management, 3rd ed., Wiley-Interscience, 2001.

[56] D. C. Hay, "Achieving buzzword compliance in Object orientation," Essential Strategies, Inc., 08 02 1999. [Online]. Available: http://www.ihs.gov/Misc/links_gateway/download.cfm?doc_id=138&app_dir_id=4&doc_file=ieoo.pdf. [Accessed 08 10 2011].

[57] DoDAF Working Group, "DoD Architecture Framework 1.5 Volume 2, 15 August 2003.," 15 08 2003. [Online]. Available: http://cio-nii.defense.gov/docs/DoDAF_Volume_II.pdf. [Accessed 10 04 2009].

[58] A. Hecht and A. Simmons, "Integrating Automated Structured Analysis and Design with Ada Programming Support Environments," NASA, 1986. [Online]. Available: http://ntrs.nasa.gov/archive/nasa/casi.ntrs.nasa.gov/19890006956_1989006956.pdf. [Accessed 08 10 2011].

[59] T. DeMarco, Structured Analysis and System Specification, New York: Yourdon Press, 1978.

[60] A. Kossiakoff and W. N. Sweet, Systems Engineering: Principles and Practices, New York: Wiley, 2003.

[61] NOAA, "NDE Project Management - NDE Context Diagram," NPOESS Data Exploitation web site, 08 07 2008. [Online]. Available: http://projects.osd.noaa.gov/NDE/proj_context_diagram.htm. [Accessed 08 10 2011].

[62] U.S. Department of Transportation, "Data Integration Glossary," 08 2001. [Online]. Available: http://knowledge.fhwa.dot.gov/tam/aashto.nsf/All+Documents/4825476B2B5C687285256B1F00544258/$FILE/DIGloss.pdf. [Accessed 08 10 2011].

[63] TechTarget, "What is a data dictionary?," SearchSOA, 02 1998. [Online]. Available: http://searchsoa.techtarget.com/definition/data-dictionary. [Accessed 08 10 2011].

[64] AHIMA e-HIM Work Group on EHR Data Content, "Guidelines for Developing a Data Dictionary," *Journal of AHIMA,* vol. 77, no. 2, pp. 64A-D, 2006.

[65] W. Stevens, G. Myers and L. Constantine, "Structured Design," *IBM Systems Journal,* vol. 32, no. 2, pp. 115-139, 1974.

[66] J. Azzolini, "Introduction to Systems Engineering Practices. July 2000," 07 2000. [Online]. Available: http://ses.gsfc.nasa.gov/ses_data_2000/000712_Azzolini.ppt. [Accessed 08 10 2011].

[67] US IRS, "Configuration Management," IRS Resources Part 2. Information Technology Chapter 27, 14 11 2008. [Online]. Available: http://www.irs.gov/irm/part2/ch17s01.html. [Accessed 08 10 2011].

[68]   J. Martin and C. L. McClure, Structured Techniques: The Basis for Case, Prentice Hall, 1988.

[69]   D. Wolber, "Structure Charts: Supplementary Notes Structure Charts and Bottom-up Implementation: Java Version," 8 12 1998. [Online]. Available: http://www.usfca.edu/~wolberd/cs112/SupplementalNotes/structureChart.doc. [Accessed 08 10 2011].

[70]   M. Page-Jones, The Practical Guide to Structured Systems Design, New York: Yourdon Press, 1980.

[71]   B. Belkhouche and J. Urban, "Direct Implementation of Abstract Data Types from Abstract Specifications," *IEEE Transactions on Software Engineering*, pp. 549-661, 1986.

[72]   P. D. Bruza and T. P. Van der Weide, The Semantics of Data Flow Diagrams, University of Nijmegen, 1993.

[73]   D. Martin and G. Estrin, "Models of Commutations and Systems— Evaluation of Vertex Probabilities in Graphical Models of Computation," *J. ACM*, vol. 14, no. 2, p. 28 ff, 1967.

[74]   C. Gane and T. Sarson, Structured Systems Analysis: Tools and Techniques, McDonnell Douglas Systems Integration Company, 1977.

[75]   SmartDraw, LLC , "What are Data Flow Diagrams?," SmatDraw Tutorials, 2011. [Online]. Available: http://www.smartdraw.com/resources/tutorials/data-flow-diagrams/. [Accessed 08 10 2011].

[76]   R. Wieringa, "A survey of structured and object-oriented software specification methods and techniques," *ACM Comput. Surv.*, vol. 30, no. 4, pp. 459-527, 1998.

[77]   B. Henderson-Sellers and J. Edwards, "The object-oriented systems life cycle," *Commun. ACM*, vol. 33, no. 9, pp. 142-159, 1990.

[78]   M. Jackson, Principles of Program Design, Academic Press, 1975.

[79]   N. Ourusoff, "Using Jackson Structured Programming (JSP) and Jackson Workbench to Teach Program Design .. Retrieved 2008-02-18," Informing Science, 18 02 2003. [Online]. Available: http://www.informingscience.org/proceedings/IS2003Proceedings/docs/091Ourus.pdf. [Accessed 09 10 2011].

[80]   K. T. Orr, "Structured programming in the 1980s," in *Proceedings of the ACM 1980 Annual Conference ACM '80*, New York, 1980.

[81]   J. D. Warnier, Logical Construction of Programs, New York: Van Nostrand Reinhold, 1974.

[82]   K. Sorensen and J. Verelst, "On the conversion of program specifications into pseudo code using Jackson structured programming," *Journal of Computing and Information Technology,* vol. 9, no. 1, pp. 71-80, 2001.

[83]   P. P.-s. Chen, "The Entity-Relationship Model: Toward a Unified View of Data," *ACM Transactions on Database Systems,* vol. 1, no. 1, pp. 9-36, March 1976.

[84]   A. P. G. Brown, "Modelling a Real-World System and Designing a Schema to Represent It," in *Data Base Description*, Douque and Nijssen, Eds., North-Holland, 1975.

[85]   P. Beynon-Davies, Database Systems, Houndmills, Basingstoke, UK: Palgrave, 2004.

[86]   H. Tardieu, A. Rochfeld and R. Colletti, La methode MERISE: Principes et outils, Editions d'Organisation, 1983.

[87]   R. B. Elmasri and N. Shamkant, Fundamentals of Database Systems, 3rd ed., Menlo Park, CA: Addison-Wesley, 2000.

[88]   "ER 2004," in *23rd International Conference on Conceptual Modeling*, Shanghai, China, 2004.

[89]   I. Feinerer, "A Formal Treatment of UML Class Diagrams as an Efficient Method for Configuration Management," der Technischen Universit¨at Wien von, Wien, M¨arz, 2007.

[90]   J. Dullea, I.-Y. Song and I. Lamprou, "An analysis of structural validity in entity-relationship modeling," *Data & Knowledge Engineering,* vol. 47, p. 167–205, 2003.

[91]   S. Hartmann, "Reasoning about participation constraints and Chen's constraints," in *Fourteenth Australian Database Conference*, Adelaide, Austria, 2003.

[92]   P. P.-S. Chen, "The Entity-Relationship Model-Toward a Unified View of Data," *ACM Transactions on Database Systems,* vol. 1, no. 1, pp. 9-36, 1976.

[93]   W. Kent, Data and Reality, 1st Book Library, 2000.

[94]   J. R. Abrial, Data semantics, Université scientifique et médicale, 1974.

[95]   R. Stamper, Information in Business and Administrative Systems, London: C. Tinling & Co. Ltd.

[96]   M. A. Jackson, System Development, Prentice-Hall, 1983.

[97]   R. Elmasri and S. B. Navathe, Fundamentals of Database Systems, 4th ed., Addison Wesley, 2003.

[98] H. Helbig and I. Glockner, "The Role of Intensional and Extensional Interpretation in Semantic Representations – The Intensional and Preextensional Layers in MultiNet," *Intelligent Information and Communication Systems,* 2006.

[99] D. C. Hay and M. J. Lynott, "UML as a Data Modeling Notation, Part 2," The Data Administration Newsletter, 01 10 2008. [Online]. Available: http://www.tdan.com/view-articles/8589. [Accessed 09 10 2011].

[100] IDEF, " IDEF1X Data Modeling Method," IDEF Iintegrated DEFinition Methods, 2010. [Online]. Available: http://www.idef.com/IDEF1x.htm. [Accessed 09 10 2011].

[101] C. Kohlhardt, "Gliffy March 2007 NewsLetter," Gliffy, 01 03 2007. [Online]. Available: http://www.gliffy.com/blog/2007/03/01/march-newsletter/. [Accessed 09 10 2011].

[102] B.-J. Hommes, The Evaluation of Business Process Modeling Techniques, TU Delft, 2004.

[103] "Software AG," 2011. [Online]. Available: http://www.softwareag.com/corporate/default.asp#idsWelcome. [Accessed 09 10 2011].

[104] "ADONIS," The BOC Group, 2011. [Online]. Available: http://www.boc-group.com/. [Accessed 09 10 2011].

[105] "Mavim," 2011. [Online]. Available: http://www.mavim.com.au/. [Accessed 09 10 2011].

[106] A. W. Scheer, ARIS. Vom Geschäftsprozess zum Anwendungssystem, Springer, 2002.

[107] A. T. e. al, "EPC Workflow Model to WIFA Model Conversion," in *2006 IEEE International Conference on Systems, Man, and Cybernetics,* Taipei, Taiwan, 2006.

[108] W. v. d. Aalst, "Formalization and Verification of Event-driven Process Chains," *Information & Software Technology,* vol. 41, no. 10, pp. 639-650, 1999.

[109] Kees van Hee, et al, "Colored Petri Nets to Verify Extended Event-Driven Process Chains," in *Proc. of the 4th Workshop on Modelling, Simulation, Verification and Validation of Enterprise Information Systems (MSVVEIS06)*, Paphos, Cyprus, 2006.

[110] E. Kindler, "On the Semantics of EPCs: A Framework for Resolving the Vicious Circle," University of Paderborn, Paderborn, Germany, 2006.

[111] M. C. Yatco, "Joint Application Design/Development," University of Missouri-St. Louis, 06 02 1999. [Online]. Available: http://www.umsl.edu/~sauterv/analysis/JAD.html. [Accessed 09 10 2011].

[112] R. Soltys and A. Crawford, "JAD for business plans and designs," 06 02 1999. [Online]. Available: http://www.thefacilitator.com/htdocs/article11.html. [Accessed 09 10 2011].

[113] J. Spencer, "DSDM (Dynamic Systems Development Method) and TOGAF (The Open Group Architecture Framework)," DSDM Consortium, 2003.

[114] K. Beck and et al, "Manifesto for Agile Software Development," 2001. [Online]. Available: http://agilemanifesto.org/. [Accessed 16 10 2011].

[115] "About DSDM Consortium," DSDM Consortium, 2011. [Online]. Available: http://www.dsdm.org/about-2. [Accessed 14 10 2011].

[116] "DSDM Atern," DSDM Consortium, 2011. [Online]. Available: http://www.dsdm.org/dsdm-atern. [Accessed 15 10 2011].

[117] DSDM Consortium, DSDM Atern Handbook V2, A. Craddock, B. Fazackerley, S. Messenger, B. Roberts and J. Stapleton, Eds., Whitehourse Press Limited, 2008.

[118] K. Barron and et al, DSDM Public Version 4.2, DSDM Consortium, 2008.

[119] C. A. Petri and W. Reisig, "Petri net," *Scholarpedia,* vol. 3, no. 4, p. 6477, 2008.

[120] J. L. Peterson, Petri Net Theory and the Modeling of Systems, Prentice Hall, 1981.

[121] J. Desel and G. Juhás, "What Is a Petri Net? Informal Answers for the Informed Reader," *Lecture Notes in Computer Science,* vol. 2128, pp. 1-25, 2001.

[122] J. Esparza and M. Nielsen, "Decidability issues for Petri nets - a survey," Bulletin of the EATCS, 1995. [Online]. Available: http://citeseer.ist.psu.edu/226920.html. [Accessed 09 10 2011].

[123] R. Lipton, "The Reachability Problem Requires Exponential Space," Yale University, 1976.

[124] P. Küngas, "Petri Net Reachability Checking Is Polynomial with Optimal Abstraction Hierarchies," in *Proceedings of the 6th International Symposium on Abstraction, Reformulation and Approximation, SARA 2005*, Airth Castle, Scotland, UK, 2005.

[125] T. Murata, "Petri Nets: Properties, Analysis and Applications," *Proceedings of the IEEE,* vol. 77, no. 4, 1989.

[126] G. Rozenberg, Advances in Petri Nets, Springer-Verlag, 1993.

[127] "Petri Nets," [Online]. Available: http://www.techfak.uni-bielefeld.de/~mchen/BioPNML/Intro/pnfaq.html. [Accessed 09 10 2011].

[128] R. David and H. Alla, Discrete, continuous, and hybrid Petri Nets, Springer, 2005.

[129] T. Araki and T. Kasami, "Some Decision Problems Related to the Reachability Problem for Petri Nets," *Theoretical Computer Science,* vol. 3, no. 1, pp. 85-104, 1977.

[130] C. Dufourd, A. Finkel and P. Schnoebelen, "Reset Nets Between Decidability and Undecidability," in *Proceedings of the 25th International Colloquium on Automata Languages and Programming, LNCS*, 1998.

[131] K. Jensen, ""Very Brief Introduction to CP-nets," Department of Computer Science, University of Aarhus, 10 04 2011. [Online]. Available: http://cs.au.dk/CPnets/. [Accessed 09 10 2011].

[132] E. P. Dawis, J. F. Dawis and W.-P. Koo, "Architecture of Computer-based Systems using Dualistic Petri Nets. Systems, Man, and Cybernetics," in *2001 IEEE International Conference on Volume 3*, 2001.

[133] E. P. Dawis, "Architecture of an SS7 Protocol Stack on a Broadband Switch Platform using Dualistic Petri Nets," in *Communications, Computers and signal Processing, 2001. PACRIM. 2001 IEEE Pacific Rim Conference on Volume 1*, 2001.

[134] A. Mazurkiewicz, "Introduction to Trace Theory," in *The Book of Traces*, V. Diekert and G. Rozenberg, Eds., Singapore, World Scientific, 1995, pp. 3-67.

[135] G. Winskel and M. Nielsen, "Models for Concurrency," in *Handbook of Logic and the Foundations of Computer Science*, vol. 4, OUP, pp. 1-148.

[136] D. Braun, J. Sivils, A. Shapiro and J. Versteegh, "State Diagrams," Object Oriented Analysis and Design Team, 2000. [Online]. Available: http://atlas.kennesaw.edu/~dbraun/csis4650/A&D/UML_tutorial/state.htm. [Accessed 10 10 2011].

[137] T. Booth, Sequential Machines and Automata Theory, New York: John Wiley and Sons, 1967.

[138] J. Hopcroft and J. Ullman, Introduction to Automata Theory, Languages, and Computation, Reading Mass: Addison-Wesley Publishing Company, 1979.

[139] E. J. McClusky, Introduction to the Theory of Switching Circuits, McGraw-Hill, 1965.

[140] D. Harel, "Statecharts: A visual formalism for complex systems," *Science of Computer Programming*, vol. 8, no. 3, p. 231–274, 1987.

[141] G. Hamon and J. Rushby, "An Operational Semantics for Stateflow.," in *Fundamental Approaches to Software Engineering (FASE)*, Barcelona, Spain, Springer-Verlag, 2004.

[142] A. Tiwari, "Formal Semantics and Analysis Methods for Simulink Stateflow Models," Unpublished, 2002.

[143] G. Hamon, "A Denotational Semantics for Stateflow," *International Conference on Embedded Software*, p. 164–172, 2005.

[144] D. Harel, "A Visual Formalism for Complex Systems.," *Science of Computer Programming*, vol. 8, p. 231–274, 1987.

[145] R. Alur, A. Kanade, S. Ramesh and K. C. Shashidhar, "Symbolic analysis for improving simulation coverage of Simulink/Stateflow models," in *Internation Conference on Embedded Software*, Atlanta, GA, 2008.

[146] M. Samek, Practical UML Statecharts in C/C++, Second Edition: Event-Driven Programming for Embedded Systems, Newnes, 2008.

[147] J. Rossberg and R. Redler, Pro Scalable .NET 2.0 Application Designs, 2005.

[148] S. M. Richard, "Geologic Concept Modeling," U.S. Geological Survey Open-File Report 99-386, 1999.

[149] T. Halpin, "Object Role Modeling: An Overview," Microsoft MDSN Library, 11 2001. [Online]. Available: http://msdn.microsoft.com/en-us/library/aa290383(v=vs.71).aspx. [Accessed 15 10 2011].

[150] "The ORM Foundation Homepage," 2010. [Online]. Available: http://www.ormfoundation.org/. [Accessed 10 10 2011].

[151] T. Halpin, "ORM 2," in *On the Move to Meaningful Internet Systems 2005: OTM 2005 Workshops*, Cyprus, 2005.

[152] T. Halpin, K. Evans, P. Hallock and B. Maclean, Database Modeling with Microsoft Visio for Enterprise Architects, Morgan Kaufmann, 2003.

[153] "DogmaModeler: Software Features," DogmaModeler, 2011. [Online]. Available:

http://www.jarrar.info/Dogmamodeler/index.htm. [Accessed 10 10 2011].

[154] "DogmaModeler Project," SourceForge, 2011. [Online]. Available: http://sourceforge.net/projects/dogmamodeler/. [Accessed 10 10 2011].

[155] "Research issues in STARLab," VUB Star.Lab, 2011. [Online]. Available: http://www.starlab.vub.ac.be/website/research. [Accessed 10 10 2011].

[156] "Tools," VUB Star.Lab, 2011. [Online]. Available: http://www.starlab.vub.ac.be/website/tools. [Accessed 10 10 2011].

[157] "Facts speak for themselves," The CaseTalk website, 2011. [Online]. Available: http://www.casetalk.com/php/. [Accessed 10 10 2011].

[158] "Infagon - Introduction," Infagon, 2011. [Online]. Available: http://www.infagon.com/. [Accessed 10 10 2011].

[159] G. Nijssen, "Doctool and CogNIAM (CogNIAM tools)," PNA Group, 2010. [Online]. Available: http://www.pna-group.com/getdoc/59792d2a-dfc2-4c02-863a-1886b7815881/Bedrijfsprofiel.aspx. [Accessed 10 10 2011].

[160] "ActiveFacts," Data Constellation, 2008. [Online]. Available: http://dataconstellation.com/ActiveFacts/index.shtml. [Accessed 10 10 2011].

[161] "DogmaStudio," VUB Star.Lab, 2011. [Online]. Available: http://www.starlab.vub.ac.be/website/tools. [Accessed 10 10 2011].

[162] "Orthogonal Toolbox," 2009. [Online]. Available: http://www.orthogonalsoftware.com/products.html. [Accessed 10 10 2011].

[163] T. Halpin and T. Morgan, Information Modeling and Relational Databases: From Conceptual Analysis to Logical Design, 2nd ed., Morgan Kaufmann, 2008.

[164] "Object-Role Modeling (ORM)," SourceForge, 2011. [Online]. Available: http://sourceforge.net/projects/orm/. [Accessed 10 10 2011].

[165] "The PLiX Project at SourceForge," SourceForge, 2011. [Online]. Available: http://sourceforge.net/projects/plix/. [Accessed 10 10 2011].

[166] FOLDOC, "Unified Modeling Language," 03 01 2002. [Online]. Available: http://foldoc.org/UML. [Accessed 10 10 2011].

[167] G. Booch, I. Jacobson and J. Rumbaugh, "OMG Unified Modeling Language Specification, Version 1.3," March 2000. [Online]. Available: http://www.uml.org/. [Accessed 10 10 2011].

[168] OMG.org, "Documents associated with UML Version 2.0," OMG, 07 2005. [Online]. Available: http://www.omg.org/spec/UML/2.0/. [Accessed 10 10 2011].

[169] OMG.org, "Documents associated with UML Version 1.3," OMG, 09 2001. [Online]. Available: http://www.omg.org/spec/UML/1.3/. [Accessed 10 10 2011].

[170] S. Mishra, "Visual Modeling & Unified Modeling Language (UML) : Introduction to UML," Rational Software Corporation, 1997. [Online]. Available: http://www2.informatik.hu-berlin.de/~hs/Lehre/2004-WS_SWQS/20050107_Ex_UML.ppt. [Accessed 10 10 2011].

[171] OMG.org, "UML Specification version 1.1," OMG, 11 08 1997. [Online]. Available: http://www.omg.org/cgi-bin/doc?ad/97-08-11. [Accessed 10 10 2011].

[172] OMG.org, "Unified Modeling Language™ (UML®)," OMG, 03 2011. [Online]. Available: http://www.omg.org/spec/UML/. [Accessed 10 10 2011].

[173] OMG.org, "Documents associated with UML Version 2.4 - Beta 2," OMG, 03 2011. [Online]. Available: http://www.omg.org/spec/UML/2.4/. [Accessed 10 10 2011].

[174] OMG.org, "Catalog of OMG Modeling and Metadata Specifications," OMG, 27 09 2011. [Online]. Available: http://www.omg.org/technology/documents/modeling_spec_catalog.htm. [Accessed 10 10 2011].

[175] J. Hunt, The Unified Process for Practitioners: Object-oriented Design, UML and Java, Springer, 2000.

[176] J. Holt, UML for Systems Engineering: Watching the Wheels, Institution of Electrical Engineers (IET), 2004.

[177] OMG.org, "UML Superstructure Specification Version 2.2," OMG, 02 2009. [Online]. Available: http://www.omg.org/spec/UML/2.2. [Accessed 10 10 2011].

[178] ". Jacobson, Interviewee, *Ivar Jacobson on UML, MDA, and the future of methodologies*. [Interview]. 24 10 2006.

[179] A. Bell, "Death by UML Fever," *Queue*, vol. 2, no. 1, 2004.

[180] G. Génova, J. Llorens, P. Metz, R. Prieto-Díaz and H. Astudillo, "Open Issues in Industrial Use Case Modeling," *Journal of Object Technology*, vol. 4, no. 6, pp. 7-14, 2004.

[181]   C. Kobryn, "Will UML 2.0 Be Agile or Awkward?," *Communications of the ACM*, vol. 45, no. 1, pp. 107-110, 2002.

[182]   F. J. L. Martínez and A. T. Álvarez, "A Precise Approach for the Analysis of the UML Models Consistency," *Perspectives in Conceptual Modeling ER 2005 Workshops AOIS BPUML CoMoGIS eCOMO and QoIS Klagenfurt Austria*, vol. 3770, pp. 74-84, 2005.

[183]   OMG.org, "Issues for UML 2.4 Superstructure and Infrastructure Revision Task Force," OMG, 2011. [Online]. Available: http://www.omg.org/issues/uml2-rtf.open.html. [Accessed 10 10 2011].

[184]   "UML FAQ," UML Forum, 2011. [Online]. Available: http://www.uml-forum.com/FAQ.htm. [Accessed 10 10 2011].

[185]   H. Tardieu, A. Rochfeld and R. Colletti, La methode MERISE: Principes et outils, 1983.

[186]   R. Skrentner, "Useless Cases and Other Irrational Artifacts," 29 08 2008. [Online]. Available: http://mydlc.com/pmi-mn/PRES/2005PDD_Skrentner.pdf. [Accessed 10 10 2011].

[187]   B. Meyer, "UML: The Positive Spin," *American Programmer*, no. Special UML, 1997.

[188]   B. Henderson-Sellers and C. Gonzalez-Perez, "Uses and Abuses of the Stereotype Mechanism in UML 1.x and 2.0," in *Model Driven Engineering Languages and Systems*, Berlin / Heidelberg, Springer, 2006.

[189]   A. Gemino and Y. Wand, "A framework for empirical evaluation of conceptual modeling techniques," *Requirements Eng*, vol. 9, pp. 248-260, 2004.

[190]   A. Gemino and Y. Wand, "Evaluation of modeling techniques based on models of learning," *Community ACM*, vol. 46, pp. 79-84, 2004.

[191]   L. Wittgenstein, Tractatus logico-philosophicus, Englewood Cliffs, NJ: Routledge Kegan Paul, 1981.

[192]   U. Frank and M. Prasse, "A framework for evaluating object oriented modeling languages: Exemplified by the examples of OML and UML," *Arbeitsberichte des Instituts für Wirtschaftsinformatik der Universität Koblenz-Landau*, vol. 7, 1997.

[193]   D. Firesmith, B. Henderson-Sellers, I. Graham and M. Page-Jones, "OPEN modeling language (OML). Reference manual," 08 12 1996. [Online]. Available: http://www.csse.swin.edu.au/OPEN/comn.html. [Accessed 10 10 2011].

[194]   U. Frank, "Evaluating modelling languages: Relevant issues, epistemological challenges and a preliminary research framework," *Arbeitsberichte des Instituts für Wirtschaftsinformatik der Universität Koblenz-Landau/Germany,* 1998.

[195]   U. Frank and C. Lange, "E-MEMO: A method to support the development of customized electronic commerce systems," *Information Systems and e-Business Management,* 2005.

[196]   J. Mylopoulos, "Conceptual modeling and Telos," in *Conceptual Modeling, Databases, and Case* , New York, John Wiley & Sons Inc., 1992, pp. 49-68.

[197]   M. Goodland and K. Riha, "SSADM - an Inroductiion," 04 03 2009. [Online]. Available: http://www.dcs.bbk.ac.uk/~steve/1/sld005.htm. [Accessed 06 11 2011].

Index

(

(min, max)-notation 121

"

"circle-box" notation 202

A

Abrial, Jean-Raymond ... 117, 121, 124, 202
accepting state 193, 194
Activity diagram..... 135, 225, 226, 239, 241
Agile development 38, 145
algorithms..... 5, 7, 8, 9, 13, 14, 53, 56, 65, 66, 198
Analysis of Alternatives............. 67
Application Domain Modeling 21
architecture. 7, 27, 37, 57, 79, 82, 83, 105, 147, 156, 184, 221, 230, 253
Arctic Tern 145, 169
ARIS 122, 129, 135
assessment...... xii, 3, 9, 14, 15, 20, 55, 60, 63, 65, 139
assumptions. 5, 7, 8, 9, 13, 15, 56, 254, 255
asymmetric choice net............. 185
Atern 145, 147, 148, 151
attribute........... 113, 114, 121, 202
automata theory 188, 198

B

Bachman diagram 115, 123
Bachman notation 121
Bachman, Charles....................... 123
Barker's notation............ 121, 125
Bayesian belief network..... 53, 65
Behavior diagrams................ 225

bijectivity 121
Block Diagram............................... 74
Booch method 216, 219
Booch, Grady. 216, 217, 218, 219, 236
Booth, Taylor..................... 192, 198
Boundedness............................. 180
Business Case 148
Business Process Model and Notation 171, 186
business process modeling... 127, 134
Business System Options ... 36, 90

C

C4ISR............................. 4, 237, 249
cardinality 115, 116, 121, 123, 125
Carnap, Rudolf 124
CaseTalk............................. 206, 210
CATWOE 44
causal influence matrix 57, 59, 67
causal loop diagram 66
Checkland, Peter..... 24, 30, 41, 43, 44, 46, 48, 49, 50
Chen, Peter Pin-San. 70, 110, 111, 113, 115, 116, 117, 118, 123, 125, 126
Class diagram 221, 222, 238, 239, 240
COBOL......................... 95, 105, 106
Cohesion 79
Communication diagram........ 228
complexity theory..................... 187
Component diagram 221, 223
Composite structure diagram 222, 224

275

Computer Aided Software Engineering..... 30, 72, 96, 105, 122, 123, 125, 140, 143, 232
computer program...................... 79
conceptual model xi, xii, 1, 2, 3, 4, 5, 8, 9, 11, 13, 14, 15, 20, 21, 22, 45, 46, 50, 54, 56, 57, 58, 59, 247, 248, 249, 250, 251, 252, 253, 254, 255, 256
conceptual modeling...ii, xi, 4, 15, 23, 24, 25, 26, 109, 127, 171, 201, 247, 248, 249
configuration system.................. 78
Constantine, Larry. 30, 70, 71, 77, 86
Context diagram..................... 33, 74
Context Diagram..... 74, 75, 85, 91, 92
control flow..... 99, 129, 131, 132, 133
Control theory........................... 188
correlation .57, 58, 59, 60, 61, 62, 67
Coupling.. 79
covering problems.................... 188
critical success factors............. 139
Crow's Foot notation..... 119, 121, 122

D

dataxii, 1, 6, 7, 8, 9, 11, 12, 13, 14, 25, 26, 27, 31, 32, 33, 37, 39, 53, 56, 63, 69, 73, 74, 77, 78, 80, 81, 82, 83, 85, 86, 87, 88, 89, 90, 95, 96, 98, 106, 107, 110, 111, 117, 118, 123, 125, 126, 129, 131, 133, 134, 141, 153, 165, 188, 198, 201, 205, 206, 228, 230, 236, 238, 240, 251, 252

Data Catalog............................ 35, 37
data dictionary.........74, 75, 76, 79
data flow 88, 89, 90, 92
data flow diagram........................24
Data Flow Diagram33, 35, 39, 71, 74, 75, 77, 80, 81, 85, 86, 87, 90, 91, 92
data flow diagrams69, 70, 74, 78, 87, 92
Data Flow Diagrams....................80
Data Flow Modeling.......23, 24, 31
data flows 91
data history..................................8
data model....................19, 33, 87
data model diagrams 70
data modeling...................123, 216
data store.............................88, 90
data streams 96
data structure............................ 103
Data Structure Diagram.. 98, 107, 117
data structuring......................... 202
database...... 12, 25, 26, 37, 39, 75, 76, 80, 82, 83, 88, 109, 110, 111, 114, 122, 123, 201, 202, 206, 209, 212, 213
database dictionary...........*See* data dictionary
database management systems ....*See* DBMS
database schemas...................... 216
databases74, 80, 81, 83, 124, 212
data-driven..1
DBMS....................... 12, 81, 83, 201
deliverables... 141, 142, 146, 152, 155, 156, 157, 158, 159, 161, 170
Department of Defense Architecture Framework......*See* DoDAF

Deployment diagram...... 222, 223
Design of Experiments 54, 57, 58, 59, 60, 66
Design Prototype............. 157, 158
developer ...xi, 1, 5, 12, 22, 50, 55, 77, 86, 118, 170, 232, 252
Devgems Data Modeler............ 122
DeZign for Databases............... 122
DFM.........*See* Data Flow Modeling
Dijkstra, Edsger Wybe 71, 96, 106, 107
directed graph 135, 186, 187, 192
Discrete Event Simulation 53, 54, 58, 59
Discrete Event System Simulation................................ 21
distributed. 1, 5, 7, 13, 80, 82, 96, 171, 173, 216, 218
Distributed Interactive Simulation................................ 10
Document Flow Analysis 93
Document Flow Diagram.......... 93
DoDAF 237, 238
DogmaModeler................. 207, 208
Double Helix...................... 142, 144
Dualistic Petri Nets 183, 184
Dynamic Systems Development Method..... 25, 29, 38, 137, 145, 146, 147, 148, 151, 152, 153, 154, 155, 156, 157, 159, 168, 170
Dynamic Systems Development Method Consortium............. 147
Dynamic view 220

E

Enterprise resource planning 135
entity..... 7, 8, 9, 13, 14, 15, 22, 31, 39, 41, 54, 58, 88, 90, 110, 113, 114, 116, 117, 118, 119, 121, 125, 206, 233, 236
Entity Life History......................... 35
Entity Relationship Diagram...35, 39, 110, 237
Entity Relationship Modeling. 23, 24, 39, 109, 110, 114, 117, 118, 123, 125
entity set............................. 119, 121
Entity-Relationship Diagram 114, 117, 122, 141
EPC.......*See* Event-Driven Process Chain
epistemology...............50, 117, 256
ER/Studio.................................. 122
Event....................................... 130
Event-Driven Process Chain.... 25, 127, 129, 130, 131, 132, 133
Event-Driven Process Chain Diagram......................... 128, 129
Evolutionary Algorithms............ 53
EXPRESS............. 25, 110, 121, 125
Extended free choice................ 185
Extreme Programming.. 146, 152, 169

F

factorial design 59, 60, 67
FDMS....................... 8, 12, 13, 22
feasibility 32, 67, 152, 155, 206
feasibility study........................... 32
federated database..................... 83
federates.................................. 5, 13
final state................................ 194
flow relation 173, 175
flowchart....77, 85, 106, 127, 196, 197
free choice net............................ 185
Function.. 131

functional decomposition. 72, 79, 81
Functional Model.. 153, 156, 157, 159, 161, 162, 165, 166
functional primitives.................. 74
Functional Prototype 156, 157
functional requirements 82
Fuzzy Logic.................................. 53
fuzzy sets..................................... 53

G

generalized nondeterministic finite state machine.............. 193
Graphviz 122

H

Halpin, Terry..................... 203, 214
Harel statechart.......................... 195
Harel, David................................. 198
hierarchy 51, 72, 79, 90, 182
High Level Architecture 22
high-level design......................... 79

I

IDEF...................... 72, 73, 82
IDEF1X 25, 82, 109, 110, 121, 125
implementation..xii, 7, 13, 30, 32, 34, 36, 59, 79, 82, 105, 114, 127, 139, 141, 142, 152, 153, 158, 160, 164, 206, 207, 212, 216, 232, 237
Infagon 206, 211
InfoDesigner................................ 207
InfoModeler.................................. 207
Information Requirements Analysis............................ 48, 49
Initial Capabilities Document.. 67
input places................ 172, 173, 174
Integration Definition .... *See* IDEF
Interaction diagrams.................. 228

Interaction overview diagram 228, 229
ISO 9000...................................... 147

J

Jackson Structured Programming 30, 71, 95, 96, 98, 99, 100, 101, 105, 106
Jackson, Michael A. 30, 71, 95, 96, 97, 98, 105, 118, 124
JAD ..*See* Joint Application Design
Java 96, 105, 216
Joint Application Design. 25, 137, 138, 142, 143
Joint Application Development 23, 138

K

KCSL Jackson Workbench...... 105
Key Performance Parameters.. 67

L

Level 1 Diagram........................... 91
life cycle.. 3, 25, 67, 137, 206, 216
Linear temporal logic 187
logical connector.............. 131, 133
Logical Data Flow Diagram....... 89
logical data model.....35, 111, 123
logical data structure....33, 34, 37
logical design.......................... 29, 36
logical process model................. 37
logical relationship.......... 131, 132
Logico-linguistic modeling 50

M

Machine Learning........................ 53
marked graph............................. 185
marking 172, 174
Markov chain..................... 183, 189
Markov property...................... 189

Martin notation 121, 125
Mealy machine 193
Meersman, Robert 202
Merise 116, 121, 125, 233
metadata 14, 76, 81, 83
meta-data model 161, 162
Meta-Object Facility 230, 231, 233
Meyer, Bertrand 231, 234
mission space 4, 5, 6, 9, 14, 56, 249
MODAF 237, 238
module specification 79
Moore machine 193
MoSCoW prioritization .. 145, 169
MSCO 12
Multiview 49
MySQL 122, 212
MySQL Workbench 122

N

NASA 2, 21
Natural language Information Analysis Methodology .... *See* NIAM
Natural ORM Architect .... *See* NORMA
Navathe, Shamkant B. 124
Neural Networks 53
NIAM 203, 206
Nijssen, Sjir 202, 203, 214
nondeterministic 26, 27, 171, 173, 183, 187, 198
nondeterministic finite state machine 193
NORMA.. 203, 204, 206, 207, 212, 213

O

Object Constraint Language ... 219

Object diagram 222, 224
Object Model 18, 22
Object Modeling Group ... 27, 216, 217, 218, 219, 230, 233, 238
Object Role Modeling 23, 26, 201, 202, 203, 204, 205, 206, 207, 208, 209, 210, 211, 212, 213
Object-modeling technique 22, 216, 218, 219, 220, 236
object-oriented 27, 73, 82, 95, 105, 202, 212, 215, 216, 217, 219, 232, 251
object-oriented analysis 216
Object-oriented analysis and design 236
object-oriented design 216
object-oriented programming. 81
object-oriented software engineering 216, 217
ontological analysis 201
ontology 50, 212
operation .... xii, 10, 17, 34, 96, 97, 99, 100, 214
Operations Research 51
Oracle Data Modeler 122
Organization unit ... 131, 133, 134
ORM .... *See* Object Role Modeling
ORM2 .... 203, 204

P

Package diagram 222, 225
parameter 2, 13, 14, 15, 125
parsing 106
Perl 96, 105
Petri net ... 26, 171, 172, 173, 174, 175, 176, 177, 178, 179, 180, 181, 182, 183, 184, 185, 198
Petri net graph 173, 174
Petri, Carl Adam 171
Physical Data Flow Diagram 89

physical design 29, 37
PowerDesigner 122, 234
Process path 133, 134
Processing Specification 35
Profile diagram 222
Program Structure Diagram ... 98, 99, 107
programmer 79, 99, 100, 103
prototype ... 89, 94, 140, 156, 157, 158, 159, 160, 163, 164, 165, 166, 167, 168
pseudo-code 74, 79

Q

QL Developer Data Modeler ... 122

R

radar 2, 5, 7, 15, 16, 17, 18, 114
Rapid Application Development 29, 146, 151, 170
Rational Rose 122, 234
Rational Software Architect ... 234
Rational Software Corporation 122, 152, 216, 217, 220, 234
Rational Unified Process 152, 220, 236
reachability 175, 177, 178, 180, 182, 183, 187
reachability graph 175, 177, 179, 180, 183
reachable markings 175, 176, 180
relational model 111, 123
relationship set 114, 115, 119, 121
representation 1, 4, 5, 6, 9, 11, 14, 15, 24, 39, 77, 81, 85, 91, 107, 109, 110, 117, 123, 186, 192, 220, 238
representative system 2

requirements xi, xii, 1, 2, 3, 5, 6, 8, 9, 11, 14, 15, 20, 21, 22, 29, 31, 32, 33, 34, 35, 36, 37, 39, 55, 69, 70, 74, 75, 82, 110, 111, 137, 138, 142, 156, 157, 158, 160, 163, 164, 165, 166, 167, 168, 169, 178, 184, 206, 249, 250, 251, 252
Requirements Catalog 35
Resource Flow Analysis 93
results validation xii
Rumbaugh, James 216, 236
run-length 100, 101, 102, 103

S

Scheer, Wilhelm-August 129
second-generation ORM .... *See* ORM2
semantic model 116, 118, 233
semantic stability 202
semantic tableau 187
semantics 12, 114, 116, 125, 130, 171, 174, 192, 196, 198, 218, 219, 233, 234, 235, 250
semiotics 117, 124
Senko, Michael 202
Sequence diagram .. 228, 230, 239
Service-Oriented Architectures 73
simulation. 2, ii, xi, xii, 1, 2, 3, 4, 6, 7, 8, 9, 10, 11, 12, 13, 14, 15, 20, 21, 22, 54, 55, 56, 57, 58, 59, 64, 82, 249, 250
simulation concept 5, 6
simulation context 4, 5, 12
simulation design .... xii, 2
simulation element 7
simulation elements ... 2, 4, 5, 6, 7, 14, 15, 20, 56
simulation space 4, 6, 7, 56, 249

SOAs ...... 82, *See* Service-Oriented Architectures
Soft Modeling 53
Soft System Methodology . 24, 30, 41, 43, 44, 45, 46, 47, 48, 49, 50
specifications. xii, 2, 3, 27, 37, 69, 74, 79, 137, 138, 219, 249
SQL 80, 122, 209, 213
SQL Maestro 122
SSADM .. 23, 24, 27, 29, 30, 31, 32, 35, 38, 49, 70, 71, 78, 87, 121
stakeholders 2, 9, 43, 44, 45, 135, 149, 150, 156, 169, 253
Stamper, Ronald 118, 124
State Diagram . 26, 191, 192, 195, 196, 197, 198
state machine .. 27, 185, 188, 189, 192, 193, 195, 196, 197, 198, 220
State Transition Modeling 23, 26, 191
Static view 220
Structure Chart. 70, 71, 78, 79, 81
Structure diagrams 221
Structured Analysis 30, 69, 70, 71, 72, 73, 79
structured data 72
Structured Design .. 30, 69, 71, 79, 86, 218
structured query language ...... *See* SQL
Structured Systems Analysis & Design Method .... *See* SSADM
Subject Matter Expert .. 54, 55, 57
subsystem 7, 15
subsystem, ... 87
syllogism 46, 51
syntactics 124
System Architect 122

system boundary 85, 89, 91
System context diagrams .... *See* Context Diagram
system development 3, 4, 70, 138, 237
systems analysis 11, 32, 47, 69, 71, 87, 123, 142, 201
Systems Engineering ... xiv, 24, 42
Systems Thinking 44

T

tableau method 178, 187
tactical ballistic missile 114
Test-driven development 170
Theater BMD 61
Timeboxing 170
Timing diagrams 228
Toad Data Modeler 122
top-down design 79
transaction data 87, 94
transition function 188, 193
Truth Table 196
Turing, Alan 188

U

UML state machine diagram. 225, 227
Unified Modeling Language 9, 23, 27, 73, 82, 112, 116, 118, 121, 125, 130, 171, 195, 199, 215, 216, 217, 218, 219, 220, 221, 226, 230, 231, 232, 233, 234, 235, 236, 237, 238, 239, 240, 243, 251, 256
Universe of Discourse 202
Use case diagram 16, 225, 227

V

V&V *See* verification and validation

verification and validation.xi, 9, 14, 59, 62
Visio 96, 122, 129, 206, 207
Visio for Enterprise Architects. 207, 209
VisioModeler 207, 208, 209
Visual Studio 207, 212, 234

W

Warnier/Orr diagram.......... 71, 106

X

XML 212, 213, 231, 232

www.ingramcontent.com/pod-product-compliance
Lightning Source LLC
Chambersburg PA
CBHW031825170526
45157CB00001B/183